Medical and Biologic Effects of Environmental Pollutants

VANADIUM

*Committee on
Biologic Effects of
Atmospheric Pollutants*

DIVISION OF MEDICAL SCIENCES
NATIONAL RESEARCH COUNCIL

NATIONAL ACADEMY
OF SCIENCES
WASHINGTON, D.C.   1974

Other volumes in the Medical and Biologic Effects of Environmental Pollutants
series (formerly named Biologic Effects of Atmospheric Pollutants):

ASBESTOS (ISBN 0-309-01927-3)   MANGANESE (ISBN 0-309-02143-X)
FLUORIDES (ISBN 0-309-01922-2)   CHROMIUM (ISBN 0-309-02217-7)
LEAD (ISBN 0-309-01941-9)
PARTICULATE POLYCYCLIC ORGANIC MATTER (ISBN 0-309-02027-1)

NOTICE: The project that is the subject of this report was approved by the Governing Board
of the National Research Council, acting in behalf of the National Academy of Sciences. Such
approval reflects the Board's judgment that the project is of national importance and appropriate with respect to both the purposes and the resources of the National Research Council.

The members of the committee selected to undertake this project and prepare this report
were chosen for recognized scholarly competence and with due consideration for the balance
of disciplines appropriate to the project. Responsibility for the detailed aspects of this report
rests with that committee.

Each report issuing from a study committee of the National Research Council is reviewed
by an independent group of qualified persons according to procedures established and monitored by the Report Review Committee of the National Academy of Sciences. Distribution of
the report is approved, by the President of the Academy, upon satisfactory completion of the
review process.

The work on which this publication is based was performed pursuant to Contract No. 68-02-0542
with the Environmental Protection Agency.

Library of Congress Cataloging in Publication Data

National Research Council. Committee on Biologic Effects of Atmospheric Pollutants.
    Vanadium.

    (Medical and biologic effects of environmental pollutants)
    Bibliography: p.
    1. Vanadium—Environmental aspects. 2. Vanadium—Physiological effect. I. Title.
[DNLM: 1. Air pollution. 2. Vanadium—Poisoning. WA754 N278v 1973]
QH545.V3N37   614.7'1   74-7131
ISBN 0-309-02218-5

Available from
Printing and Publishing Office, National Academy of Sciences
2101 Constitution Avenue, Washington, D.C. 20418

Printed in the United States of America

PANEL ON VANADIUM

RICHARD U. BYERRUM, College of Natural Science, Michigan State University, East Lansing, Michigan, *Chairman*

ROBERT E. ECKARDT, Medical Research Division, Esso Research and Engineering Company, Linden, New Jersey

LEON L. HOPKINS, JR., Agricultural Research Service, Fort Collins, Colorado

JOSEPH F. LIBSCH, Whitaker Laboratory, Lehigh University, Bethlehem, Pennsylvania

WILLIAM ROSTOKER, Department of Materials Engineering, University of Illinois, Chicago, Illinois

CARL ZENZ, Medical College of Wisconsin, Milwaukee, Wisconsin

WILLIAM A. GORDON, Spectrochemical Analysis Unit, Lewis Research Center, National Aeronautics and Space Administration, Cleveland, Ohio, *Consultant*

JOHN T. MOUNTAIN, Cincinnati, Ohio, *Consultant*

SAMUEL P. HICKS, Department of Pathology, University of Michigan Medical Center, Ann Arbor, Michigan, *Associate Editor*

T. D. BOAZ, JR., Division of Medical Sciences, National Research Council, Washington, D.C., *Staff Officer*

iii

Acknowledgments

This document was written by the Panel on Vanadium under the chairmanship of Dr. Richard U. Byerrum. Each section was prepared initially by a member of the Panel, but subsequent combinations and transpositions have blurred the exact boundaries of authorship in a few instances. However, the total document was reviewed and approved by the Panel and represents the combined effort and cooperation of its members. The section on the industrial aspects of vanadium and its compounds was primarily the responsibility of Dr. William Rostoker. Dr. Robert E. Eckardt prepared the section on vanadium in the air, as well as the appendix on the desulfurization of oil. Dr. Joseph F. Libsch contributed the section on vanadium from other sources and in the environment in general, with Dr. Leon L. Hopkins, Jr., responsible for the section on food. The part dealing with biologic effects was written principally by Dr. Carl Zenz, with some portions by Dr. Byerrum. The material on sampling and analysis, which appears as Appendix B, is the contribution of Mr. William A. Gordon.

The preparation of the document was assisted by the review by members of the Committee on Biologic Effects of Atmospheric Pollutants*

---

*The Committee's name has since been changed to Committee on Medical and Biologic Effects of Environmental Pollutants; its purpose and scope remain the same.

and by the many useful comments offered by anonymous reviewers responding to Dr. Samuel P. Hicks, who served as Associate Editor. The Panel is greatly indebted to the Editor of the Division of Medical Sciences, Mr. Norman Grossblatt, whose efforts resulted in the formation of a single coherent report from the contributions of several authors.

Dr. Robert J. M. Horton of the Environmental Protection Agency (EPA) was of invaluable assistance with his wise counsel and his knowledge of reference material. In addition, he helped in obtaining important documents and some translations. Dr. Kenneth Bridbord served as liaison officer for EPA. Informational assistance was provided by the National Research Council Advisory Center on Toxicology, the National Academy of Sciences Library, the National Library of Medicine, the National Agricultural Library, the Library of Congress, and the Air Pollution Technical Information Center.

Acknowledgment is made of the assistance given by the Environmental Studies Board, National Academy of Sciences–National Academy of Engineering (NAS-NAE), and the divisions of the National Research Council.

Contents

Introduction

Vanadium compounds were recognized as industrial hazards more than a half-century ago. In areas where vanadium compounds were used industrially, breathing of particulate material with a high vanadium content was observed to cause an intense, dry cough accompanied by irritation of the nose, eyes, and throat. Prolonged industrial exposure to vanadium compounds was found to result in emaciation, anorexia, nausea, and diarrhea.

Widespread interest in vanadium toxicity in animals and man has been spurred by the recent discovery of high concentrations of vanadium compounds in the fly ash from combustion of residual petroleum products and coal from some parts of the world and by the finding of vanadium in many plant, animal, and human tissues.

Vanadium in trace amounts has recently been shown to be nutritionally essential for rats and chicks, and it has been known for some time to be a blood constituent of several marine organisms.

The data related to the effects of environmental vanadium compounds on plant, animal, and human life are fragmentary. This report attempts to collate and evaluate these data.

1

Industrial Processes
Involving Vanadium
and Its Compounds

CHEMISTRY OF VANADIUM AND ITS COMPOUNDS

Pure vanadium is not a common commercial metal, but it has potentially useful attributes. The metal, when of appropriate purity, is ductile at room temperature and above. It has been used in all the common wrought forms—sheet, wire, and tube. Vanadium melts at about 1900 C, which is at least 400 C higher than the melting points of steels. Because of its reactivity with both oxygen and nitrogen and with the oxide materials used for refractories, it must be melted in vacuum with cold-mold, arc-fusion furnaces. Even at its high melting point, the vapor pressure is low (estimated boiling point, 3350 C); and, because the melting system is closed, there is no expectation of release in any form to the atmosphere.

Pure vanadium is very resistant to corrosion[156,157] in simple aerated saline solution—at least as resistant as stainless steels and probably as resistant as titanium. It has good resistance to dilute halide acids, but suffers on exposure to concentrated solutions. It is inferior to stainless steels in resistance to strong alkaline solutions.

At high temperatures, vanadium metal resists oxidation up to the melting point of vanadium pentoxide (690 C). Above that temperature, the oxidation rate of the metal becomes catastrophic. Even a low concentration (a few percent) of vanadium in alloys decreases their resis-

3

tance to oxidation; this is a problem particularly with steels. The addition of vanadium to steels that normally form tight, adherent, and protective oxides converts the scale (oxide) into a porous, bulky, and friable substance that readily falls off the steel surface. If vanadium oxide as a constituent of the fly ash of burned fuel comes into contact with the hot surface of steel, it decreases the oxidation resistance of the steel. The abnormally short lifetime of boilers that burn vanadium-bearing fuels has long been an unsolved technical problem. Vanadium pentoxide combines readily with other metal oxides to form vanadates. In some ways that are not understood, the presence of vanadates or their mode of formation prevents oxide films from growing as a dense, adherent, and protective coating. Therefore, if vanadium-bearing fuels are used, the solid effluents will contain oxides of other metals in the steam-generating system.

Ferrovanadium, mainly an alloy of iron and vanadium, is one of the common commercial forms in which vanadium is added in steelmaking. It is brittle but not friable. It is not as corrosion-resistant as pure vanadium, but sufficiently to resist rusting in the normal open storage of a steel plant. Aluminothermic vanadium, an alloy of aluminum and vanadium, is used in titanium-alloy production. Its mechanical properties are similar to those of ferrovanadium, and it might be even more corrosion-resistant. Vanadium carbide, used in the iron and steel industry, is also very corrosion-resistant.

Various chemical forms of vanadium are widely used in the chemical industry as catalysts. The physical properties of the important chemical species are given in Table 1-1. Commonly, there is finite solubility in water. The boiling points of vanadium chlorides and oxychlorides are low.

TABLE 1-1   Some Physical Properties of Important Vanadium Compounds[a]

Compound	Melting Point, C	Boiling Point, C	Solubility in Water, g/100 cc	
			Cold Water	Hot Water
Vanadium pentoxide	690	1750[b]	0.07[c]	0.07[c]
Sodium metavanadate	630	no data	21.1	38.8
Vanadium tetrachloride	(below room temperature)	148.5	soluble[b]	no data
Vanadium oxychloride	(below room temperature)	126.7	soluble[b]	no data
Ammonium vanadate	200[b]	no data	0.52	6.95[b]

[a]Data from Weast,[205] unless otherwise indicated.
[b]Decomposes.
[c]Data from Seidell.[165]

The effect of vanadium pentoxide on the degradation of oxidation resistance of steel was once ascribed to a high vapor pressure. That is not true. Its ability to attack other metals at a distance reflects a very low surface tension in the molten state. [187]

Vanadium can form covalent bonds to yield organometallic compounds.[96] The known number of such compounds is increasing steadily. Some of these, such as vanadium acetylacetate, are finding use in polymerization processes and are likely to appear as inclusions in the polymer product.

VANADIUM MINERALS AND MINERAL TREATMENT

There are no naturally occurring, highly concentrated forms of vanadium minerals. This is despite the fact that there are at least 50 distinguishable mineral species[44] and that, in terms of the available earth's crust, vanadium is about as plentiful as zinc and nickel. A few commercial deposits (of carnotite and of titaniferous magnetite) contain as much as 3% vanadium pentoxide; most are much leaner, generally containing 0.1–1% vanadium pentoxide.

Apart from vanadium sulfides of Peru, which are no longer in production, commercially exploited vanadium is in some oxidized form, usually as a metal vanadate. Almost invariably, the extraction of vanadium is associated with some other valuable product. Thus, the carnotite ores can be processed only when the market for uranium is advantageous, and the exploitation of vanadium from titaniferous magnetite is contingent on the market for magnetite concentrates and on transportation costs.

In the United States, vanadium can be extracted from the following mineral resources: carnotite, a uranium–vanadium mineral, $K(UO_2)_2(VO_4)_2$; phosphate-rock deposits; titaniferous magnetite associated with vanadium; and vanadiferous clays.

Vanadium also occurs in small amounts in fossil fuels. Fly-ash derivatives from combustion have been only irregularly processed for vanadium recovery. The vanadium as vanadium pentoxide in the fly ash shows up in the air-dispersed effluent.

Comminuted mineral awaiting extractive treatment is often stockpiled in the open, where it is subject to weathering by air and rain. However, the vanadium is locked in a highly oxidized and distributed form. It is not known whether surface drainage picks up soluble vanadium, which would then become a water pollutant.

The situation is different if the vanadium is in a partially processed state. It is common, in handling vanadium-bearing minerals, to convert the vanadium and sometimes the other minerals to a water-soluble form. The conversion process involves salt-roasting, in which the vanadium-bearing material in finely comminuted form is mixed with 6–10% sodium chloride[156] and heated (roasted) to about 850 C. Under these circumstances, oxidized vanadium is converted to sodium metavanadate, $NaVO_3$, which is highly water-soluble. The hot-water leaching process is about 90% efficient. The pregnant liquor is made to precipitate sodium metavanadate by addition of acid until the pH is 1–3. When dried and fused, the product is sold as commercially pure vanadium pentoxide.

The solid residues from the leaching operation will contain some water-soluble materials. In the case of magnetite pellets, this presents no problem, because these are later fed to the blast furnace, where the vanadium emerges in insoluble oxidized form in the slag or perhaps as refractory carbide in the pig iron. In any event, further dispersion cannot occur. The salt-roasting of the ferrophosphorous product from phosphate-rock smelting should be similarly safe, in that the residue from the leaching operation is returned via a smelting operation that locks the residual vanadium into inert form.

However, the residues from roasting and leaching of carnotite ores, vanadiferous clays, and vanadium-enriched slags are likely to be heaped on the ground or perhaps used as landfill, and they will be subject to rain and groundwater drainage. Although this has not been discussed in the literature, it is reasonable to expect these tailings dumps to be a source of water pollution for many miles around.

Vanadium ores are generally not imported into the United States. Because of the low concentration of vanadium in the native ore, treatment to increase the concentration is applied near the mining site. Vanadium is brought into this country as commercial oxide (fused sodium metavanadate); this is a friable material, and dust from packaging and handling might present a hazard.

2

Industrial Uses of
Vanadium and Its Compounds

METALLURGIC USES

The use of constructional steels that contain about 0.1% vanadium is substantial and increasing[132] (see Table 2-1). There are also many traditional tool and die steels that contain vanadium in concentrations of 0.2–3%,[132] but these are produced in relatively small amounts. In all cases, vanadium is added either as ferrovanadium (an iron–vanadium alloy containing 50–80% vanadium) or as vanadium carbides (VC and V_2C) at a stage in melt refining such that the alloy is efficiently dissolved in the liquid steel. Economics demands that as little as possible be lost to the slag or furnace refractories. In conventional steel-smelting operations, there seems little basis for expecting much loss of vanadium in fume or airborne dust. The metal dissolves readily in the steel and has negligible vapor pressure. The oxide dissolves readily in the slag, which is a reasonably inert glassy substance. The oxide, either combined or uncombined, has negligible vapor pressure at the operating temperatures. Although the slag may be used as landfill and subjected to rain and drainage, there is no evidence that traces of vanadium could exist in any fugitive form.

A large part of all the steel made finds its way back to steel-smelting operations as scrap. In the refining stage of such operations, the contained alloy is oxidized out of the liquid metal as completely as feasible.

7

TABLE 2-1   Vanadium Content of Constructional Steels[a]

Steel	Vanadium Content, %	Typical Uses
C–V	0.10–0.20	Railroad equipment—shafts, arms, connecting rods, driving axles, crank pins, guides, piston rods—parts for heavy machinery, pumps, and diesel engines
Cr–V	0.05–0.50	Low-carbon: carbonized gears, camshafts, piston pins, pressure vessels, tubing, stampings
		Medium-carbon: axles, shafts, steering arms, gears, springs, arbors, spindles, shafting, cyanided gears, bolts, washers, forgings
		High-carbon: tools, ball bearings, wearing plates
Mn–V	0.05–0.20	Heavy forgings, plates for transportation equipment and for tanks, rivets, bolts
V–spring	0.08–0.2	Heavy and light springs for transportation equipment, heavy machinery, and instruments

[a]Derived from National Academy of Sciences.[132]

In the slowly declining use of the open-hearth furnace, vanadium from the scrap steel would simply be oxidized into the slag, which would then be run off, granulated, and discarded with reasonable safety.

Today, a very high proportion of scrap steel is recycled by way of the basic-oxygen furnace (BOF), which refines steel with a high-pressure jet of pure oxygen. This injection produces enormous clouds of fume that consists essentially of metallic oxides—primarily iron oxides. However, some portion of the vanadium (and all other alloying elements) in the scrap charge might be expected to be incorporated in the fume. Clouds of fume discharged from the plant deposit particles on the surrounding land for many miles.

As defined by Patty,[138] fume represents "solid particles generated by condensation from the gaseous state, generally after volatilization from melted substances and often accompanied by chemical reaction such as oxidation." Fume can usefully be characterized as a suspension of particles less than 0.01 μm in diameter. Some measure of the problem can be seen in this statement: "During 1968 steel produced in the BOF was 48 million short tons and the estimated vanadium content of the particulate emissions was 0.02 percent. The emission factor for the BOF has been estimated at 46 pounds of particulate per ton of steel and the degree of emission control at 97 percent."[53] Using these figures, one can calculate a total fume emission of about 221 tons of vanadium per year, of which 7 tons escape to the atmosphere. Compared with nationwide emissions from coal and oil,[53] this may seem small, but the intensity of deposition is high because the discharge is highly concentrated and local (see Chapter 3).

The high-cost tool and die steels of today are produced largely in vacuum-induction and vacuum arc-melting systems, which are closed and do not vent to the outside.

The ferrovanadium used for alloy additions in steelmaking is produced by smelting in electric-arc furnaces. The charge consists of scrap steel, fused sodium metavanadate, and carbon, with silicon, aluminum, or a combination of the two as reducing agent. The furnace is open to the atmosphere. Most of the charge exists as metal slag or carbonaceous gases, but a substantial amount evolves as fume. "The particle size generally ranges from 0.1 to 1 μm and the total particulate emission is 240 pounds per ton of charge"; "during 1968 vanadium emissions to the atmosphere resulting from the production of ferrovanadium totaled 144 short tons."[53] This emission constitutes a potentially serious problem in the vicinity of ferrovanadium-producing plants, of which there are few in the United States.

Vanadium carbides as alloy additions are produced by the carbothermic reduction of sodium metavanadate in a closed system operating under vacuum. This is a very clean operation, in that little vanadium escapes into the atmosphere.

Whatever melting operation is used, steel ingots are processed by metal-deformation operations to finished or semifinished plate, bars, and forgings. Most metal working is done while the workpiece is hot— 870–1200 C. At these temperatures, steel oxidizes superficially, and thick oxide or mill scale is formed. Oxide or mill scale is recycled to the melting operations, except for the pickling step. In preparation for cold finishing, steels are pickled. In pickling, the surface of the steel is dissolved in baths of strong acids. While surface defects are thereby removed, so also is a considerable volume of metal. The spent acids contain water-soluble salts of all the alloying elements (including vanadium). Unless treated properly, the large quantities of "spent" acids could be a potential source of pollution.

Apart from its use in steels, vanadium is a major alloying element in high-strength titanium alloys. The alloy of greatest production for many years has contained 4% vanadium. Because of their reactive nature, these alloys are melted under vacuum in a closed system—a very clean operation. During the processing of ingots to wrought shapes, there is scale formation, and various grinding and machining operations produce comminuted materials. Most of these are not recycled, but the materials are quite inert and do not become airborne. Pickling of plate, sheet, and forgings is common and necessary. Both strong acids in aqueous solution and fused salts (primarily sodium hydroxide) are used. Sludge and spent acids are usually mixed to neutralize each other, but the residue is rich in halide salts of vanadium.

CHEMICAL USES

Such vanadium compounds as vanadium oxychloride, vanadium tetra-chloride, and vanadium trisacetylacetonate are used as catalysts in poly-merization processes to produce soluble copolymers derived from ethylene and propylene. There is also substantial use in the manufacture of synthetic rubber of the ethylene–propylene diene types.[132] Because the polymer product in the reactor is a viscous liquid, the entrapment may reach 500 ppm, but more often it is 50–100 ppm. There may be a hazard related to the extraction of these materials from plastics used for food packaging and pharmaceutical and medical applications.

CONTAMINATION BY VANADIUM FROM METAL PRODUCTS AND USES

As already described, by far the greatest tonnage of vanadium products is associated with the metallurgic industry. Metallurgic processing involves raw materials (various ores), intended products (such as ferrovanadium, vanadium, steel alloys, and titanium alloys), and waste products (flue gases and slag). Each aspect of processing may contribute to the eventual distribution of vanadium in air and in water.

Metallurgic products containing vanadium, particularly steel and titanium alloys, are themselves often processed—e.g., by flame cutting, welding, and brazing for joining. Processing operations at high temperatures are of particular interest, because they may involve considerable smoke and fume.

A potential source of vanadium for air and water distribution involves metallurgic processing industries that are not specifically concerned with producing vanadium compounds, but involve vanadium-containing waste or byproducts.

This section is particularly concerned with potential concentrations of vanadium from the waste products of industrial operations, from the processing of metals that contain vanadium, and from metallurgic processes in which vanadium compounds are a byproduct or waste product.

Industrial Processing

Naturally occurring world resources of vanadium by country, ore type, vanadium pentoxide content, etc., are given in a report by the National Materials Advisory Board.[132(pp.8–9)] The concentration of vanadium in these ores is relatively low—about 0.1–2.0%. In addition, vanadium-rich products result from industrial processing, providing sources of vanadium that do not occur in nature.

Vanadium residues from fly ash, slags, and spent vanadium catalysts are apparently being used in the production of vanadium products in the United States and elsewhere. These sources generally have a considerably higher vanadium concentration than naturally occurring ones and therefore account for a greater concern in handling, controlling point sources, and preventing general contamination. Ash from petroleum-fired facilities, for example, may contain up to 70% vanadium pentoxide,[118] although the concentration varies significantly, depending on the vanadium content of the fuel oil. Vanadium pentoxide concentrations of 5–18% have been reported for fly-ash residues commonly processed. Although reported as vanadium pentoxide, the vanadium present in fly ash is rarely in this form and most probably exists in one of several reduced forms—as vanadium tetroxide, trioxide, and dioxide—but not necessarily in the form of simple vanadium oxides.

Vanadium-rich slags also represent a potential source of vanadium. The Nuremberg plant of Gesellschaft für Elektrometallurgie, for example, has recently installed facilities to process the slag from Highveld Steel and Vanadium Corporation, Ltd. (a member company of Anglo American Corporation of South Africa).[63] Processing of a vanadiferous magnetite ore reportedly provides a vanadium pentoxide concentration of 20–25% in the slag. Faulkner Hudson[60(p.104)] reports processing of a high-vanadium slag from the Domnarvret Iron Works in Falun, Sweden, the vanadium being present in the trivalent form. He also cites[60(p.112)] Archibald concerning the vanadium concentration in slags from steel-making converters that use a charge of relatively high vanadium content and operate especially for recovery of vanadium; he indicates that the vanadium content there may reach 10%, or even higher. However, in the United States, neither the slag from processing of vanadiferous magnetite deposits nor the richer slag (up to 2.3% vanadium pentoxide) of steel plants that produce vanadium tool steels is now an attractive source.

Another potential source is spent vanadium-containing catalysts, which also contain very high concentrations and are being processed in the United States. Thus, vanadium residues—whether fly-ash, slag, or spent-catalyst—appear to be significant potential vanadium sources in processing ferrovanadium or other vanadium products.

Waste Products

Waste products of metallurgic operations normally include gases and slag. Published information on the air distribution of vanadium from a point source (e.g., a manufacturing facility) is very scarce.

Table 2-2 provides information on the atmospheric concentrations of vanadium at the boundary of a typical steel plant over a period of approximately 9 months in 1967. It is apparent that the average concentration corresponded with the average of 13 Pennsylvania cities.

The available information indicates that vanadium air contamination by manufacturing plants may be less than that from power production and heating facilities that use high-vanadium fossil fuels. Table 4-7, (p. 38) also suggests that industry is not the major source of vanadium in the air. In contrast, it appears that the rapid expansion of heat and power production using residual oils and coal is a primary consideration.

There is little information concerning the possible distribution of vanadium contained in slag from metallurgic processing. It is apparent that the slag produced as waste products from metallurgic operations may contain substantial concentrations of vanadium, although the concentrations in normal steel-plant operations (representing the prepon-

TABLE 2-2   Atmospheric Concentrations of Vanadium at Boundary of Typical Steel Plant, 1967[a]

Date Collected	Vanadium Concentration, $\mu g/m^3$
Jan. 1	0.054
Feb. 1	0.063
Feb. 22	0.040
Apr. 30	0.090
May 16	0.082
May 31	0.105
June 7	0.098
June 29	0.040
Aug. 3	0.040
Aug. 18	0.107
Sept. 8	0.092
Sept. 21	0.081
Oct. 5	0.040
Average (range)	0.072 (0.040–0.107)
Average of 13 Pennsylvania cities for 1967	0.072

[a]Data from N. E. Whitman, Bethlehem Steel Corporation (personal communication). All samples were collected at the plant boundary in a community, but at an elevation of about 50 ft. The Staplex high-volume air sampler was used, with 8 X 10-in. fiberglass filters, for the sample collection. Air concentrations were calculated as vanadium, rather than vanadium pentoxide.

derant volume) are much less. The chemical form of the vanadium in such slag is not known, but the slag is considered relatively insoluble. It therefore does not appear to represent a significant source of vanadium distribution by the leaching action of surface water and rainwater. However, any concentrations of salt-roasted vanadium ore or residues do represent a potential source of distribution if exposed to rain or surface water, inasmuch as salt-roasting converts the ore to a soluble form.

Processing of Metals Containing Vanadium

The common ferrous alloys containing vanadium are fabricated for use in many shapes and thus require processing. Such processing includes chemical treatment (such as pickling of scale) and high-temperature operations (such as welding, flame cutting, and hot working).

Table 2-3 shows the concentrations of vanadium pentoxide in the breathing zone of workers concerned with four vanadium metal operations: during the furnace addition of vanadium, during the tap of a heat, in oxyacetylene cutting, and in arc-welding beams. The data are reassuring with respect to minimal vanadium fume exposure in these operations, which would be expected to produce the greatest source of fume.

In addition to potential fume containing vanadium associated with the welding of vanadium-containing steels, it has been reported that the coating of some welding rods contains vanadium and produces metal fume containing vanadium.[180] Unfortunately, little information exists on the toxic substances present as impurities and on their concentrations in arc-welding fumes.

Other data from a single steel plant include information pertaining to acid wastes from pickling operations and baghouse dust.

Acid wastes from pickling operations, for example, indicated a vanadium concentration of 0.02% in hydrochloric acid, nitric acid–hydrofluoric acid, and sulfuric acid tanks, although the solid sludge in the bottoms of these tanks contained 0.24% vanadium in the hydrochloric acid tank and 0.16% in the nitric acid–hydrofluoric acid tanks. The vanadium content of the baghouse dust in the same plant over a 2-month period varied as follows: 0.06, 0.06, 0.13, 0.24, and 0.05%.

It is common to process the water from all mill operations in water-treatment plants, often in such a way as to precipitate the metals in solution with sodium hydroxide. Analysis of this precipitate in one instance showed the vanadium content to be 0.07 wt %.

TABLE 2-3 Concentrations of Vanadium Pentoxide in Air at Point Sources at a Steel Plant[a]

Location	Operation	Vanadium Content of Steel, %	Concentrations of Vanadium Pentoxide in Breathing Zone, mg/m^3 of air[b]
7-ton furnace	Furnace addition of vanadium	0.15–0.25	0.006
		0.15–0.25	0.007
		0.15–0.25	0.078
		0.15–0.25	0.007
		1.8 –2.1	0.019
		1.8 –2.1	0.013
7-ton furnace	Tap of heat	0.5 –0.7	0.004
		0.5 –0.7	0.011
		0.5 –0.7	0.004
		0.5 –0.7	0.02
Press forge	Oxyacetylene cutting of forging	1.02	0.002
		1.05	0.008
		0.15–0.4	0.004
		0.15–0.4	0.015
		0.15–0.4	0.003
		0.15–0.4	0.011
		0.1 –0.15	0.01
		0.1 –0.15	0.002
		0.1 –0.15	0.005
		0.1 –0.15	0.001
Beam-yard weld bed	Arc-welding beams	0.8	0.003
		0.8	0.006
		0.8	0.002
		0.8	0.002
		0.8	0.004
		0.8	0.004
		0.8	0.003
		0.8	0.005
		0.8	0.004
		0.8	0.005

[a]Data from N. E. Whitman, Bethlehem Steel Corporation (personal communication).
[b]Threshold limit value for vanadium fume as vanadium pentoxide: 0.1 mg/m^3 in 1961.[6]

CONTAMINATION BY VANADIUM FROM NONMETALLURGIC PRODUCTS AND PROCESSING

The steel industry uses by far the greatest tonnage of vanadium, primarily as an alloying element in steels. Various vanadium compounds, generally inorganic, are also important in industrial processing. The major nonmetallurgic use is in the chemical industry, but vanadium has had and still has a number of uses in the ceramics and electronics industries.

Chemical Production

The principal uses of vanadium in the production of chemicals are as a catalyst in the synthesis of sulfuric acid and in the oxidation of numerous organic compounds to commercial products. Table 2-4 lists some of the chemicals so processed. Estimates of vanadium consumed by the chemical industry vary from 197 tons in 1965 to 132 tons in 1967,[132 (p.41)] with projections of 200–250 tons by 1976.[132 (p.41)]

The oxidation of sulfur dioxide to sulfur trioxide in the production of acid consumes by far the largest amount of vanadium (estimated at 76 tons of vanadium pentoxide in 1965).[132] Vanadium pentoxide on asbestos as a carrier, or as a bed 20–50 in. deep, supported on a screen, is reported to be the preferred catalyst for the reaction because of its long life, high average efficiency, immunity to poisoning (by arsenic, chlorine, etc.), and great physical ruggedness.[132 (p.41)]

The vanadium oxide catalyst is also used in two closely related processes—those involving the oxidation of aromatic hydrocarbons to phthalic anhydride and the oxidation of benzene to maleic anhydride. However, other vanadium compounds have been, and are, also used. For example, ammonium metavanadate is mentioned in connection with the manufacture of adipic acid, vanadium oxytrichloride in connection with the production of ethylene–propylene synthetic rubber,

TABLE 2-4   Some Processes Using Vanadium Catalysts

Production of sulfuric acid
Manufacture of phthalic anhydride
Manufacture of maleic anhydride
Production of aniline black
Oxidation of cyclohexanol to adipic acid
Oxidation of ethylene to acetaldehyde
Oxidation of anthracene to anthraquinone
Oxidation of toluene or xylene to aromatic acids
Oxidation of furfural to fumaric acid
Oxidation of hydroquinone to quinone
Oxidation of butene-2 and 1,3-butadiene to maleic anhydride
Ammonolysis/oxidation of toluene, *m*-xylene, *p*-xylene, and propylene
Preparation of vinyl acetate from ethylene
Manufacture of cyclohexylamine from cyclohexanol and ammonia
Catalytic combustion of exhaust gases
Catalytic synthesis of ethylene–propylene rubber

and vanadium trichloride in stereospecific catalyst systems and as an intermediate in making high-purity metals and intermetallics.[132]

In dye manufacture and dyeing, vanadium compounds are widely used in the production of aniline black. Vanadium salts are added as catalysts to a mixture of aniline hydrochloride and potassium or sodium chlorate.[132]

Vanadium compounds are used as mordants in the dyeing and printing of cotton and, particularly, for fixing aniline black on silk. Ammonium metavanadate has been used as a catalyst in the dyeing of leather and fur. Some modern quick-drying inks depend on the addition of ammonium metavanadate for their performance.[132]

Generally, total consumption of vanadium in these chemical processes is small; furthermore, its principal use is as a catalyst, not as an ingredient. Potential hazards other than in point-source industrial operations appear limited. But these uses of vanadium do leave two unique aspects for consideration. First, any vanadium compound mechanically entrained in the product will have very wide distribution. Second, what happens to the spent catalyst? In the production of sulfuric acid, for example, any vanadium carried over into the acid could be very widely distributed because of the extensive industrial use of sulfuric acid. Similarly, the vanadium compound in quick-drying inks, dyes, etc., could have wide distribution and, in the case of inks, possible concentration during disposition of waste material. There appears to be no information on this question; fortunately, however, any concentration of vanadium would most probably be very small.

As mentioned earlier, some spent catalysts are currently being processed to vanadium products as residue "ores." The disposition of other spent catalysts could offer a potential problem.

Polymers

Closely allied to the chemical industry is polymer synthesis and processing. Extensive research has illustrated the role of the transition metal elements, including vanadium, as effective catalysts in promoting polymerization.[65] A large number of catalyst-promoter systems have been developed, although they are not fully interchangeable with respect to a desired result, such as structure and molecular-weight distribution. In fact, the influence of a given transition metal in catalysis is fairly specific;[132 (p.45)] for example, with 1,3-diene monomers—such as butadiene or isoprene—vanadium catalysts favor a 1,4-*trans* structure in the polymer. Also, vanadium compounds are extensively used as catalyst components in the synthesis of amorphous copolymers derived from ethylene and propylene and of EPDM copolymers synthesized from ethylene, propylene, and a nonconjugated diene. It has been shown[151] that polymerization results depend on the nature of the transition metal used in the catalyst. It appears likely that vanadium will continue to be used as a catalyst in copolymerization processes, particularly in copolymerization of ethylene and propylene monomers.[16] In addition, vanadium catalysts—i.e., the oxychloride ($VOCl_3$) and others—may play a role in the polymerization of high-density polyethylene. See Table 2-4 for a list of some processes that use vanadium catalysts.

Total consumption of vanadium in polymer synthesis and processing is small, although growing rapidly; estimates show it will reach 100–150 tons by 1975.[132 (p.45)] The products of such polymers (e.g., packaging films) are widely distributed—sometimes in connection with food—and large volumes of them are incinerated. The question of potential hazard from vanadium extracted from polymers during food storage and from vanadium emission during incineration has been raised—i.e., whether any vanadium catalyst is mechanically entrained during processing. Fortunately, tests of plastics for vanadium have shown concentrations of only 1 ppb or less, which are not considered hazardous.

Ceramics

The use of vanadium in the ceramics industry appears to be decreasing. It is reported that vanadium compounds are not currently used in the manufacture of fiberglass, frits, or refractories. However, ammonium metavanadate continues to play a primary role in various ceramic glazes, particularly the zircon vanadium blues. These glazes are not used in dinnerware, but in such applications as wall tile. Furthermore, the vanadium is tightly bound in substitutional solid solution in the zircon structure and is therefore very stable.

The Electronics Industry

The growth of the electronics industry, particularly that related to solid-state devices, has emphasized the role of a number of transition elements, including vanadium. Considerable research is under way, and it is likely that new applications will evolve. For example, the insulator-to-metal transition in vanadium tetroxide at about 60 C is intriguing; the resistance drops by several orders of magnitude over a very narrow temperature range. Similarly, vanadium as vanadium pentoxide has a high absorption of ultraviolet—a useful characteristic in window glass and eyeglasses.

There appears little chance that such future applications of vanadium compounds could threaten environmental health, because the vanadium will be tightly bound. In point-source considerations (such as manufacturing facilities), however, problems could develop, as indicated by the use of europium-activated yttrium orthovanadate as a cathodoluminescent phosphor.[193] In this instance, the superior characteristics of the new phosphor, in terms of brightness and operating characteristics relative to television, depend in a specific way on the use of ammonium metavanadate. Manufacturing operations related to this phosphor developed mean point concentrations of vanadium of 0.844 mg/m³ (well above the threshold value of 0.5 mg/m³) with a consequent occupational-health problem, fortunately without permanent damage. This phosphor is no longer used. It is to be expected that the solid-state applications of vanadium will be specific, i.e., will depend on the unique characteristics of the element.

CONCLUSIONS

The natural occurrence of vanadium and its compounds and their use in industrial processes and products make it ubiquitous. In particular, extensive use of the element in catalysis provides a potential for wide distribution, but the data make it difficult to envision ambient air or water concentrations of vanadium that even approach the threshold values. Although there have been exceptions, point-source concentrations also appear to be generally well below the vanadium pentoxide threshold values of 0.5 mg/m³ for dust and 0.05 mg/m³ for fume currently cited. The handling of vanadium residues—fly ash, slag, and spent catalyst— appears to be a more urgent point-source concern.

3

Some Sources of Vanadium in the Ambient Air

All coal and all crude petroleum oils contain a variety of metallic components as impurities. One of these metals is vanadium. The concentration of vanadium in coal and crude petroleum oils is low but appears to vary rather widely among different sources of coal or petroleum and within a given coal field or a given oil field.

The chemical forms in which vanadium is present in coal and crude oil are not known as completely as might be desired. There is reason to believe, however, that a substantial portion of the vanadium is chemically associated with the organic portions of the coal and oil and therefore cannot be removed by simple processes of physical separation, such as flotation in the case of coal and water washing in the case of oil. Some studies have led to the conclusion that at least some of the vanadium may be present as porphyrin compounds (in analogy with hemoglobin, an iron porphyrin, and chlorophyll, a magnesium porphyrin).

As used in this chapter, "vanadium" in the various tables refers to elemental vanadium. This does not imply, however, that the vanadium is emitted in the form of metallic vanadium. It is highly unlikely that any of the vanadium emitted during the combustion of fossil fuels is in the metallic form. It is probably emitted in oxidized form, but not necessarily as simple oxides.

The Council on Environmental Quality[164] has estimated that some 7.2 million tons of particulate matter were emitted into the air in the United States during 1969 as a result of the combustion of fossil fuels (coal and oil). If it is assumed that the burning of coal and oil resulted in the emission of some 20,000 tons of vanadium into the air in 1969, then it can be estimated that vanadium constituted only about 0.28% of the total particulate matter emitted.

COAL

A report prepared for the Environmental Protection Agency[53] presented two estimates of vanadium emission into the ambient air in the United States in 1968 as a result of the combustion of coal: 1,750 and 3,760 tons. Both estimates were arrived at by assuming that just under 509 million tons of coal were consumed in the United States in 1968. The lower estimate was reached by assuming that all the coal consumed in the United States had an average vanadium content of 22.5 ppm, that 65% of the total vanadium in the coal appeared as fly ash (with the remaining 35% being retained in the ash staying in the furnaces), that 90% of the coal-burning units had fly-ash collection equipment installed, and that this equipment operated at 85% efficiency. The higher estimate was based on emission data from six coal-fired power plants reported by Cuffe and Gerstle.[45] In this study, emissions were determined for six coal-fired steam generators of typical design, each fired with a different type of bituminous coal (originating in Illinois, Pennsylvania, Ohio, West Virginia, and Kentucky). Samples of the fly ash were taken before and after fly-ash collection and analyzed spectrochemically for a variety of metals, including vanadium.*

The estimate based on the Cuffe and Gerstle data may not be particularly accurate and should be used with caution, for several reasons:

1. The coal burned during the tests represented only a small portion of the coal mined in the various regions of the United States; its vana-

---

*The vanadium concentrations in the fly-ash samples taken after collection ranged from 0.88 to 6.6 grains/10,000 standard cubic feet (scf) (0.201 to 1.5 mg/m^3 of flue gas, with an average of 2.91 grains/10,000 scf, 0.66 mg/m^3). On the basis of consumption of 509 million tons of coal in the United States in 1968, 90% application of control, a vanadium concentration of 2.91 grains/10,000 scf (0.66 mg/m^3), and 160 scf of flue gas per pound of coal, the vanadium emission due to the combustion of coal was estimated at 3,760 tons.

dium content may not be typical of that of all United States coal.

2. Cuffe and Gerstle reported that their trace-metal analysis on the fly-ash samples was accurate only to ± 50%.

3. It can be calculated from the Cuffe and Gerstle data that the fly ash before collection contained 16–115 lb of vanadium per million pounds of coal burned, with an average of 57 lb. If it is assumed that the fly ash constitutes only 65% of the total ash, then the total vanadium content of the ash would be 25–177 lb/1,000,000 lb of coal burned, with an average of 87 lb. A United States average of 87 ppm for bituminous coal appears too high, perhaps by a factor of 3. On the basis of the most recent Bureau of Mines data,[2] the weighted average vanadium content of coal mined in the United States is about 30 ppm.

Almost all the coal consumed in the United States contains some vanadium.[2] This conclusion is based on spectrochemical analysis of the ash from 827 samples, for vanadium and 28 other elements, of United States commercial coal used industrially. The average vanadium content of coal from the important coal-producing areas of the United States is shown in Table 3-1. According to U.S. Bureau of Mines statistics,[28] the amount of bituminous coal produced in the United States in 1969 was about 556 million tons (Table 3-2). As indicated in Table 3-2, the weighted average vanadium content of coal produced in 1969 was calculated to be 30 ppm.

The disposition of the bituminous and anthracite coal produced in the United States in 1969 is shown in Table 3-3. From these data, an estimate was prepared of the amount of vanadium emitted into the air in the United States as a result of the combustion of coal in 1969. In

TABLE 3-1 Vanadium Content of Coal Samples in the United States, by Region

Coal-Producing Region	No. Samples Analyzed	Average Vanadium Content, ppm
Eastern[a]	600	30
Interior[b]	123	34
Western[c]	104	15

[a]Alabama, eastern Kentucky, Maryland, Ohio, Pennsylvania, Tennessee, Virginia, and West Virginia.
[b]Arkansas, Illinois, Indiana, Iowa, Kansas, western Kentucky, Missouri, and Oklahoma.
[c]Arizona, California, Colorado, Idaho, Montana, New Mexico, North Dakota, Oregon, South Dakota, Texas, Utah, Washington, and Wyoming.

TABLE 3-2   Bituminous Coal Production in the United States, 1969

Coal-Producing Region	Coal Produced, million tons[a]	Fraction of Total, %	Vanadium Content ppm[b]	Vanadium Content tons
Eastern	436.4	78.5	30	13,092
Interior	92.2	16.6	34	3,135
Western	27.5	4.9	15	412
TOTAL	556.1	100.0	30	16,639

[a]Data from *Bituminous Coal Facts.*[28]
[b]Data from Abernethy *et al.*[2]

preparing the estimate, it was necessary to use a number of assumptions, including the following:

1. The vanadium content of bituminous coal averages 30 ppm.
2. The vanadium content of anthracite coal averages 125 ppm. (The data are limited; 125 ppm is the average of two values reported by Abernethy and Gibson.[1])
3. Fly ash constitutes 65% of the total coal ash (35% of the total ash remains in the furnace).
4. Collection of ash by electric-power utilities reduces vanadium content by 85%.
5. Collection of ash in manufacturing operations reduces vanadium content by 60%.

TABLE 3-3   Coal Disposition in the United States, 1969[a]

Type and Use of Coal	Amount of Coal, 1,000 tons
Bituminous	
Electric-power utilities	308,462
Manufacturing	93,248
cement mills	9,131
steel and rolling mills	5,560
other manufacturing, bunker, and mining	78,557
Retail deliveries	12,665
Coking	92,901
TOTAL	507,276
Anthracite	9,275

[a]Data from *Bituminous Coal Facts.*[28]

6. Collection of ash of the coal designated as being in the "retail deliveries" category reduces vanadium content by 50%.

7. The coking of coal results in no emission of vanadium into the air, assuming that essentially all the vanadium remains in the coke and eventually appears in the blast-furnace slag, etc., when the coke is used in metallurgic processes.

The estimates arrived at on the basis of these assumptions are shown in Table 3-4. The estimate indicates that about 1,750 tons of vanadium were emitted by burning bituminous coal in 1969 and perhaps an additional 375 tons by burning anthracite coal. The anthracite-coal estimate is open to considerably more question than the bituminous-coal estimate, because very limited data are available on either the vanadium content of anthracite coal or the degree to which fly-ash control equipment is used when anthracite coal is burned.

There is little or no information available on the chemical form of vanadium in the particulate matter emitted into the air during coal combustion. It probably occurs in oxidized forms, but they are unlikely to be simple oxides of vanadium. It is more likely that vanadium is associated chemically with one or more of the other metals contained in the coal burned. There is also a paucity of information on the range of particle sizes in which the vanadium compounds are emitted into the air.

TABLE 3-4 Estimated Emission of Vanadium Resulting from Coal Burning in United States, 1969

Type and Use of Coal	Coal, 1,000 tons	Vanadium in Coal, tons[a]	Vanadium in Fly Ash, tons[b]	Control of Fly Ash, %	Vanadium Emitted to Air, tons
Bituminous					
Electric-power utilities	308,462	9,254	6,015	85	902
Manufacturing[c]	93,248	2,797	1,818	60	727
Retail deliveries	12,665	380	247	50	124
Coking	92,901	2,787	−	100[d]	0
SUBTOTAL	507,276	15,218	8,080		1,753
Anthracite	9,275	1,159	753	50	377
TOTAL					2,130

[a]Assuming that bituminous coal contains 30 ppm vanadium, and anthracite coal, 125 ppm.
[b]Assuming that fly ash amounts to 65% of total coal ash.
[c]Includes bunkers and mining.
[d]Coking produces no fly ash; vanadium remains in coke and finally appears in blast-furnace, etc., slag.

About all that can be said is that the fly-ash particles containing vanadium probably cover a relatively broad spectrum of sizes and that the size range is influenced probably by a host of factors—including the source of the coal burned, its degree of pulverization, the type of furnace in which it is burned, the combustion conditions, and the type of fly-ash collection equipment used.

Another potential source of vanadium emission into the air is the burning coal-waste piles, or "culm banks," in the coal-producing regions of the United States. According to the Bureau of Mines,[117] nearly 300 of these coal-waste piles are currently burning. These piles emit smoke, gas, and ash. The ash probably contains vanadium and all the other metals present in the coal, but there are no known data on this point. Accordingly, the vanadium emitted into the air from this source cannot be measured.

PETROLEUM

All crude oils of petroleum origin appear to contain vanadium as an impurity. As indicated in Tables 3-5–3-7, the concentration can vary from less than 1 to as high as 1,400 ppm, depending on the source of the crude oil. Natural gas and the so-called natural-gas liquids (ethane, propane, and butane) do not contain detectable amounts of vanadium (i.e., they contain less than 0.05 ppm); accordingly, combustion of these products does not result in any significant emission of vanadium compounds into the ambient air.

TABLE 3-5   Vanadium Content of Domestic Crude Oils[a]

Source of Oil	Vanadium Content, ppm
Arkansas	9.3
California	50.0
Colorado	0.4
Kansas	15.1
Louisiana	0.5
Montana	78.0
New Mexico	0.1
Oklahoma	4.0
Texas	2.6
Utah	4.6
Wyoming	49.7

[a]Derived from W. E. Davis and Associates.[53]

TABLE 3-6   Vanadium Content of Venezuelan Crude Oils[a]

Source and Type of Oil	Vanadium Content, ppm
Western Venezuela	
Bachaquero heavy	390
Bachaquero	370
Bachaquero light	49
Barinas	165
Boscan	1,400
Cumarebo	0.7
Lagunillas heavy	300
La Rosa medium	230
Mara	220
Mototan No. 7	390
Taparito	450
Tia Juana light	100
Tia Juana medium	200
Tia Juana heavy	300
Urdaneta	430
Eastern Venezuela	
Cachipo	14
Guanipa	110
Jusepin	26
Oficina light	57
Oficina heavy	62
Pedernales	230
Pilon	510
Quiriquire	95
San Joaquin	0.6
Temblador	56
Tigre	160
Tucupita	84

[a]Derived from W. E. Davis and Associates.[53]

In the distilling and refining of crude petroleum to make fuels and other commercial products, nearly all the vanadium remains in the so-called residua, which contain the highest-molecular-weight portions of the crude oil. These residua are the principal source of the products that are sold as residual fuel oils under such designations as No. 5 and No. 6 or Bunker C fuel oil; therefore, the particulate matter released into the air during the combustion of residual fuels does contain vanadium. However, the distillate fuels made from crude petroleum—including aviation gasolines, motor gasolines, jet fuels, kerosenes, automotive diesel fuels, and home heating oil—contain either no vanadium or barely detectable traces (0.05 ppm or less). Thus, the combustion of these fuels contributes practically no vanadium to the ambient air.

TABLE 3-7   Vanadium Content of Middle East
Crude Oils[a]

Source and Type of Oil	Vanadium Content, ppm
Agha Jari (Iranian)	36
Ain Dar (Arabian)	51
Ain Zalah (Iraqi)	95
Bai Hassan (Iraqi)	19
Gach Saran (Iranian)	114
Jambur (Iraqi)	6
Kirkuk (Iraqi)	30
Kuwait	30
Qatar	3
Safaniya (Arabian)	80
Shedgum (Arabian)	18
Uthmaniyah (Arabian)	51
Wafra (Neutral Zone)	52
Zubair (Iraqi)	20

[a]Derived from W. E. Davis and Associates.[53]

Some of the vanadium in at least some crude oils is present in the
form of organic compounds (e.g., porphyrins) of very high boiling
point. Low concentrations of vanadium (2–3 ppm) have, therefore,
been observed in some heavy-gas oil fractions (obtained by vacuum
distillation of the bottoms remaining after the distillation of crude oil
in pipe stills at atmospheric pressure). The heavy-gas oils are not sold
commercially. They are used as feed stocks in other conversion and refining processes, such as catalytic cracking and hydrodesulfurization,
which remove most of the vanadium and other metals present.

In summary, the only products of petroleum origin that are potential sources of significant emission of vanadium to the ambient air are
crude oils and heavy fuel oils. (The latter may contain nearly all the
metals present in the crude oil from which they were prepared.) An
exception to this generalization, which is increasing in importance,
should be borne in mind: Processing of heavy (residual) fuel oils to
reduce their sulfur content coincidentally can significantly reduce
their vanadium content and other metallic impurities. This is illustrated in the following tabulation of data from Radford and Rigg,[145]
which shows the vanadium content of residual fuel oils made from a
representative Venezuelan crude oil as the sulfur content was pro-
gressively reduced from 2.6% to 0.5% by hydrodesulfurization of the
atmospheric residuum:

	sulfur content, %	vanadium content, ppm
undesulfurized	2.6	218
desulfurized	1.0	105
desulfurized	0.5	59

The vanadium content of the fuel oil was reduced as the sulfur content was reduced. The situation will be similar when low-sulfur residual fuel oils are made by hydrodesulfurizing heavy-gas oils obtained by vacuum distillation of atmospheric residua and then blending the desulfurized-gas oils with the undesulfurized vacuum bottoms. (Additional information on this subject is presented in Appendix A.) Thus, vanadium emission into the air resulting from the combustion of residual fuel oils should decrease as desulfurized fuel oils are supplied in response to regulations that limit the sulfur content of fuel oils. This trend will be of special importance along the eastern seaboard of the United States, because much of the residual fuel oil burned in this area is of Venezuelan origin, and an increasing amount of this fuel oil is being desulfurized to meet regulations imposed particularly by the more highly industrialized states along the East Coast.

As noted earlier, the vanadium content of crude oils varies widely. The vanadium content of residual fuel oils also varies widely, depending on a number of factors, such as the source of the crude oil from which they are prepared, the yield of residual fuel from the crude oil, and the degree to which the residual fuel was processed to meet sulfur or other specifications.

Table 3-8 indicates the vanadium content of some residual fuel oils. It can be seen that vanadium in residual fuel oils can vary from less than 1 to well over 200 ppm. Residual fuels made from Venezuelan crude oils are at the high end of the spectrum.

In a recent report prepared for the Environmental Protection Agency,[53] it was estimated that the residual fuel oils burned in the United States in 1968 contained a total of slightly under 19,000 tons of vanadium and that 90% of this total (some 17,000 tons) was emitted into the air. The estimate is shown in Table 3-9.

Note that the estimate was made by assuming vanadium contents for residual fuel oils made from United States, South American, and Middle Eastern and other crude oils. Unfortunately, firm data on the vanadium content of the residual fuel oils actually consumed in the United States are not available. But it is known that the vanadium content of residual fuel oils from South American (particularly Venezuelan) crude oils is, on the average, considerably higher than that of residual fuel oils

TABLE 3-8   Vanadium Content of Untreated
Residual Fuel Oil Samples[a]

Sample and Source of Oil	Vanadium Content, ppm
United States	
Philadelphia No. 6	78
Santa Fe No. 6	40
Port Arthur No. 6	28
Louisiana (Timbalier Bay)	3
Venezuela	
Ceuta	230
Mesa	100
Vengref Menez Bunker C	230
Vengref Menez No. 6	114
Oficina	107
Columbia (Orito)	50
Libya (Sarir)	0.5
Angola (Cabinda)	2
Iran (Agha Jari)	76
Abu Dhabi (Zakum)	2
Sumatra (Minas)	0.4

[a]Derived from W. E. Davis and Associates.[53]

from any other known crude-oil source. Accordingly, the vanadium contents assumed are at least directionally correct.

The consumption of residual fuel oil in the United States in 1968 was estimated at 668 million barrels.[53] The available data indicate that this quantity increased by 8% to 722 million barrels in 1969 and by another 11% to a total of 804 million barrels in 1970.

Table 3-10 gives a breakdown of estimated residual fuel oil consumption in the United States in 1969 and 1970 by source of crude used to

TABLE 3-9   Estimated Vanadium in Residual Fuel Oils Consumed in the United States, 1968[a]

Source of Crude Oil	Quantity of Residual Fuel Oil, million barrels	Vanadium Content ppm	Vanadium Content tons
United States	258.3	30	1,280
South America	376.0	280	17,350
Middle East and other	33.9	50	280
TOTAL	668.2	172	18,910

[a]Derived from W. E. Davis and Associates.[53]

produce the residual fuel. A range of concentration is given for the vanadium content of the residual fuel oil from each source, because they are not known exactly.

The total vanadium content of the residual fuel oil consumed in 1969 is estimated at 13,800–21,300 tons, and in 1970, at 15,700–24,200 tons. The degree of control of vanadium emission into the air during the combustion of residual fuel oil (by virtue of ash remaining in the furnaces and the application of ash collectors to stack gases) probably is around 10%.[53] With this assumption, the amount of vanadium emitted to the air as the result of the combustion of residual fuel oil is estimated at 12,400–19,200 tons in 1969 and at 14,100–21,800 tons in 1970.

Information on the chemical form in which vanadium is emitted into the air during the combustion of residual fuel oil and the size range of the vanadium compound particles emitted is limited at best. According to Bowden *et al.*,[32] the fly ash from residual fuel oil combustion may contain the following variety of vanadium compounds: V_2O_3, V_2O_4, V_2O_5, $Na_2O \cdot V_2O_5$, $2Na_2O \cdot V_2O_5$, $3Na_2O \cdot V_2O_5$, $2NiO \cdot V_2O_5$, $3NiO \cdot V_2O_5$, $Fe_2O_3 \cdot V_2O_5$, $Fe_2O_3 \cdot 2V_2O_5$, $Na_2O \cdot V_2O_4 \cdot 5V_2O_5$, and $5Na_2O \cdot V_2O_4 \cdot 11V_2O_5$. No generalizations can be drawn regarding the size of the particles emitted into the air during combustion. About all that can be said is that the particulate matter emitted into the air during the combustion of residual fuel oil covers a relatively wide spectrum of particle sizes.

If the gross concentrations of vanadium in the ambient air are considered to be a matter of concern from a health standpoint, then research programs aimed at determining the chemical and physical forms of the vanadium compounds emitted during the combustion of residual fuel oil would be appropriate.

Estimates suggest that about 31 billion gallons of residual oil were burned throughout the United States in 1970. In the same year, approximately 93 billion gallons of gasoline were consumed, and perhaps half this much of other distillate materials, such as jet fuels, kerosenes, home heating oils, and automotive diesel fuels. The vanadium content of petroleum distillate products—including motor gasolines, jet fuels, kerosenes, home heating oils (No. 2 heating oil), and automotive diesel fuels—is typically below 0.05 ppm, which is the minimal amount detectable by the analytic method used.[58] Thus, if 31 billion gallons of residual oil (vanadium content, 50 ppm or greater) were burned, and 135 billion gallons of distillate fuels, including motor gasoline (average vanadium content, less than 0.05 ppm), were burned, the contribution from motor gasoline and other distillate fuels would be insignificant in comparison with that from the residual fuels. (The vanadium content of

TABLE 3-10 Estimated Residual Fuel Oil Consumption and Vanadium Content, United States, 1969 and 1970[a]

Source of Crude Oil	Assumed Vanadium Content, ppm	1969		1970	
		Quantity of Residual Fuel Oil, million barrels	Vanadium Content, tons	Quantity of Residual Fuel Oil, million barrels	Vanadium Content, tons
United States	25–50	230.7	952–1,903	226.3	936–1,872
Canada	25–50	16.1	66–132	23.7	98–196
Other western hemisphere	200–300	381.8	12,600–18,900	438.0	14,454–21,681
North Africa	10–20	62.0	103–205	92.0	152–304
West Africa	10–20	23.4	39–77	19.7	33–66
Middle East	50–90	5.8	48–86	2.9	24–43
Far East	10–20	2.2	4–7	1.8	3–6
TOTAL		722.0	13,812–21,310	804.4	15,700–24,168
90% of total			12,431–19,179		14,130–21,751

[a]Based on data from American Petroleum Institute and other organizations.

residual fuel oil is about 1,000 times greater than that of the distillate fuels.) Accordingly, the amount of vanadium emitted to the air from mobile sources—such as aircraft, automobiles, buses, and trucks—and oil-fired home heating equipment is low, but its significance is unknown. The only substantial sources of vanadium in the fossil-fuel category are coals and the so-called residual fuel oils burned in stationary combustion devices.

WOOD, OTHER VEGETABLE MATTER, AND SOLID WASTES

According to the Council on Environmental Quality,[164] an estimated total of 12.8 million tons of particulate matter were emitted into the air in the United States during 1969, much of it as the result of the burning of wood (including forest fires), other vegetable matter (agricultural wastes), and solid wastes. Although this particulate matter contained some vanadium, it is unlikely that the total amount was particularly high or even significant, inasmuch as the limited available information indicates that the vanadium content of wood and other vegetable matter is low. Schroeder,[158] for example, has reported that the vanadium content of land plants is around 1.6 ppm, and other investigators have found much lower values.[177]

4

Vanadium in the Environment

ATMOSPHERE

Some vanadium has been found in the air over even relatively unpopulated areas of the earth. In the eastern Pacific Ocean area,[84] the range of concentrations at nine different locations between San Diego and Honolulu was 0.02–0.8 ng of vanadium per cubic meter of air, with an average concentration of 0.1 ng/m³. In a remote area of northwestern Canada,[146] the range of vanadium concentrations at five different sites was 0.21–1.9 ng/m³, with an average of 0.72 ng/m³. The natural origins of vanadium that might contribute to the observed concentrations of vanadium in such areas are marine aerosols (produced when small particles are suspended in the air by wind) and continental dust (produced by wind erosion of rocks and soil). Some vanadium presumably would be injected into the atmosphere by volcanic action, but this source would be negligible in comparison with the other two.

One method of testing whether marine aerosols and continental dust are the sources of vanadium in the air over unpopulated areas would be to predict vanadium concentrations relative to the concentrations of other elements known to arise from those sources. Zoller et al.[217] have used this technique. They have assumed that most of the airborne sodium and chloride in a marine atmosphere are present on sea-salt particles. An upper limit of the marine contribution of va-

32

nadium in a given location can be estimated by multiplying the observed aerial concentration of Na^+ and Cl^- by the vanadium:sodium and vanadium:chloride ratios in seawater. Likewise, an upper limit of the vanadium contribution from continental dust can be predicted by assuming that all the observed airborne iron arises from that source and then multiplying the air iron concentration by the vanadium:iron ratio in rock or soil. This prediction need not be restricted to the elements mentioned. For example, magnesium concentrations in marine aerosols can be multiplied by the vanadium:magnesium ratio in seawater to predict an upper limit of vanadium concentration in sea air. The concentration of vanadium relative to the concentrations of other elements from natural sources is shown in Table 4-1.

This technique of predicting vanadium concentration was used, and it was found that the concentration of vanadium in the air over windward Hawaii was higher than had been predicted on the basis that all the vanadium had come from seawater and continental dust, as shown in Table 4-2. The additional vanadium presumably came from man's activities. Results of the same sort of prediction and measurement of vanadium concentration in the atmosphere of rural Canada are shown in Table 4-3. The observed vanadium concentration was about the same as would be predicted if all the vanadium had come from continental dust. Although there is some variation in the predicted vanadium concentration in Table 4-3, depending on the element used for compari-

TABLE 4-1 Concentration Ratios of Vanadium to Other Elements in Natural Sources

Other Element	Concentration Ratio, Vanadium:Other Element				
	Diabase[a]	Crustal Average[a]	Soils[b]	Sedimentary Rocks[b]	Seawater[a]
Iron	0.0031	0.0027	0.0026	0.0039	
Manganese	0.18	0.14	0.12	0.19	
Aluminum	0.0031	0.0017	0.0014	0.0018	
Zinc	2.9	1.9	2.0	1.6	
Scandium	7.1	6.1	14.3	13	
Cobalt	4.8	5.4	12.5	5.6	
Antimony	220	676	100	100	
Lanthanum	8	4.5	2.5	3.2	
Sodium	–	–	–	–	2×10^{-7}
Magnesium	–	–	–	–	1.5×10^{-6}
Chlorine	–	–	–	–	10^{-7}

[a]Data from Mason.[115]
[b]Data from Vinogradov.[202]

TABLE 4-2   Predicted Vanadium Concentrations in Air over Windward Hawaii from Natural Sources[a]

Source	Element Used as Basis	Concentration of Element, ng/m^3	Predicted Vanadium Concentration, ng/m^3	
Marine aerosols	Sodium	5,300	0.0011	0.0010
	Magnesium	670	0.0010	
Continental dust	Iron	38	0.099	
	Manganese	0.86	0.103	0.114
	Aluminum	101	0.141	
TOTAL (predicted)[b]			0.11	

[a]Derived from Zoller et al.[217]
[b]Observed: Acid-soluble   0.24 ± 0.10
        Insoluble   0.11 ± 0.08
        Total   0.35 ± 0.13
Predicted: observed = 31%

son, it seems that man's activities contribute only negligible amounts of vanadium to the air over rural Canada.

Table 4-4 shows vanadium concentrations in the air over various rural areas of the United States. The concentration of vanadium is significantly higher in the rural atmosphere of 9 states on the eastern seaboard from Maine to South Carolina than in the other rural areas of the country. The air concentrations of vanadium in some parts of

TABLE 4-3   Predicted Vanadium Concentrations in Air over Rural Canada from Natural Sources[a]

Source	Element Used as Basis	Concentration of Element,[b] ng/m^3	Predicted Vanadium Concentration, ng/m^3	
Marine aerosols	Sodium	44	0.0000088	0.000005
	Chlorine	13	0.0000013	
Continental dust	Iron	210	0.55	
	Manganese	6.7	0.80	
	Aluminum	186	0.26	0.74
	Scandium	0.11	1.57	
	Lanthanum	0.20	0.50	
	Zinc	15.4	31	anomalous
	Antimony	0.25	25	
TOTAL (predicted)[c]			0.74	

[a]Derived from Zoller et al.[217]
[b]Data from Rahn.[146]
[c]Observed (average of five winter stations): 0.72 ± 0.50
Predicted:observed = 100%

rural United States do not differ greatly from those observed in a remote area of Canada.

Vanadium concentrations in the air over a number of urban areas are shown in Table 4-5. The areas can be separated into two groups on the basis of vanadium concentrations in the atmosphere. Group A is characterized by average vanadium concentrations ranging from below the limit of detection of the analytic method used to 22 ng/m^3, and group B, by average concentrations between 25 and 1,320 ng/m^3. Although there are some differences in concentration values reported for particular areas (probably arising from different sampling locations, atmospheric conditions during sampling, and analytic errors), the vanadium concentration is clearly much higher in the group B cities than in the group A cities. The group B areas are all in the eastern United States.

TABLE 4-4 Vanadium in Air over Nonurban Areas[a]

	Yearly Average Vanadium Concentrations, ng/m^3				
State	1965	1966	1967	1968	1969
Rhode Island	24	32	48	34	47
Vermont	14	29	64	41	53
Maine	15	11	29	4	18
Maryland	6	25	44	19	34
New York	5	7	9	6	18
North Carolina	3	6	5	5	8
South Carolina	2	3	4	3	7
New Hampshire	2	10	8	6	12
Pennsylvania	2	2	2	1	4
Mississippi	< 1	< 1	3	1	–
Texas	1	< 1	< 1	< 1	2
Arkansas	< 1	< 1	< 1	< 1	2
Oklahoma	< 1	< 1	< 1	< 1	< 1
New Mexico	6	< 1	1	–	–
Nevada	2	< 1	< 1	< 1	< 1
Indiana	2	1	1	< 1	1
Colorado	1	< 1	< 1	< 1	1
Wisconsin	1	–	1	–	1
Iowa	< 1	–	–	–	–
Missouri	< 1	< 1	1	1	< 1
Arizona	2	< 1	< 1	< 1	< 1
Nebraska	< 1	< 1	< 1	< 1	< 1
Montana	< 1	< 1	< 1	< 1	< 1
Oregon	< 1	< 1	1	< 1	< 1
Wyoming	< 1	< 1	< 1	< 1	< 1
South Dakota	< 1	< 1	3	< 1	1
California	< 1	< 1	1	< 1	< 1

[a]Data obtained from National Air Surveillance Networks and supplied by E. C. Tabor.

TABLE 4-5   Vanadium in Air over Urban Areas[a]

Area	No. Samples	Vanadium Concentration, ng/m^3 Range	Average	Reference
Group A[b]				
Honolulu	12	0.68–2.4	1.4	Hoffman[83]
Los Angeles (UCLA)	18	2.2–28	12	Hoffman[83]
San Francisco	9	1.5–11	6.2	Zoller et al.[217]
Niles, Mich.	25	1.2–16	4.5	Rahn et al.[147]
Northwestern Indiana	25	5–18	7.4	Harrison et al.[75]
Chicago	22	2–120	22	Brar et al.[34]
Urban England	?	6–60	21	Keane and Fisher[95]
Group B[b]				
Buffalo, N.Y.	6	500–2,000	800	Pillay and Thomas[143]
New York City	25	74–2,000	458	U.S. Dept. of HEW
New York City	270	90–320	170	Morrow and Brief[125]
New York City	52	310–2,790	1,320	Kneip et al.[97]
Boston	10	400–2,000	600	Zoller and Gordon[216]
Boston	12	90–1,320	494	Moyers et al.[129]
Boston	9	89–2,410	940	Gordon et al.[71]

[a]Derived from Zoller et al.[217]
[b]See text for explanation of groups.

A recent report by Blosser[30] presents data obtained with a variety of analytic methods, giving vanadium concentrations in the air over a number of United States cities, which agree well with the data presented in Table 4-5.

That emissions from manufacturing processes might not be a major contributor to airborne vanadium is suggested by the data in Table 4-6. Comparison of the average concentrations of vanadium in air samples taken during the fall and winter (first and fourth quarters) with the concentrations in samples taken during the spring and summer (second and third quarters) indicates that there was generally twice as much vanadium in the air during the colder period. These data strongly suggest that power production and heating with residual-oil products and coal are responsible for the increased vanadium concentrations in cities in the northeastern United States in winter and perhaps, to some extent, for those in rural areas.

Table 4-7 shows predictions of upper limits of vanadium concentrations based on calculations similar to those in Tables 4-2 and 4-3. The data suggest that continental dust is a much more important contributor to atmospheric vanadium in San Francisco and northwestern Indiana than in New York City. These calculations support the conclusion, drawn from the data in Table 4-6, that man's activities result in significant air contamination by vanadium in the northeastern part of the country.

Vanadium in the air occurs in particulate form, generally in particles 0.1–2 μm in diameter.[30] Zoller *et al.*[217] report that vanadium in the atmosphere of Boston is associated mainly with the smaller particles, as might be expected of particles formed in a high-temperature process, such as oil combustion, followed by condensation of the vaporized material.

EARTH'S CRUST

Vanadium occurs ubiquitously in the earth's crust. Vinogradov[202] has summarized data on the concentrations of vanadium in a variety of rocks and soils, as shown in Table 4-8. Vanadium concentrations vary according to the type of rock or soil and are higher in shales and clays than in other rocks or soils. The average concentration of vanadium in the crust is 150 mg/kg. The data shown in Table 4-8 agree well with those presented by Schroeder,[158] which indicate that the vanadium concentration is 135 mg/kg in igneous rock, 130 mg/kg in shale, and about 20 mg/kg in sandstone and limestone. Vanadium does not occur in the crust as the free metal, but rather as relatively insoluble salts, which most commonly contain vanadium in the trivalent state.

TABLE 4-6   Vanadium in Air over Areas with High Concentrations, by Season[a]

Area	Year	Vanadium Concentration, ng/m^3	
		2nd and 3rd Quarters	1st and 4th Quarters
Bridgeport, Conn.	1962	61	130
Hartford, Conn.	1964	37	110
New Haven, Conn.	1964	62	110
Baltimore, Md.	1965	48	94
Brockton, Mass.	1965	37	48
Somerville, Mass.	1962	94	230
Springfield, Mass.	1964	33	89
Bridgeton, N.J.	1965	21	23
Bethlehem, Pa.	1965	15	48
Portsmouth, Va.	1965	29	49
Richmond, Va.	1965	52	91

[a]Derived from Schroeder.[158] (Data taken from *Air Quality Data . . . 1966.*[200])

TABLE 4-7   Predicted Vanadium Concentrations in Air over Several United States
Urban Areas from Continental Dust[a]

Element Used as Basis	Concentration of Element, ng/m^3	Predicted Vanadium Concentration, ng/m^3	
San Francisco, areawide study (9 samples)[b]			
Iron	1,920	5.0	
Manganese	19	2.3 } 2.9	
Aluminum	990	1.4	
		Observed	6.2 ± 2.1
		Predicted:observed	4%
Northwestern Indiana, area-wide survey (25 samples)[c]			
Iron	3,900	10.1	
Manganese	130	15.3 } 9.5	
Aluminum	1,950	2.7	
		Observed	8.2 ± 2.8
		Predicted:observed	114%
New York City, areawide survey (~ 270 samples)[d]			
Iron	2,980	7.4	
Aluminum	2,040	2.8 } 3.4	
Silicon	4,840	1.5	
		Observed	170
		Predicted:observed	2%

[a]Derived from Zoller et al.[217]
[b]Data from Rahn et al.[147]
[c]Data from Harrison et al.[75]
[d]Data from Morrow and Brief.[125]

WATER

Although the concentration of vanadium in seawater is usually low, 2–
29 μg/liter, the total amount in the oceans is about 7.5 × 10^{12} kg.[115]
Inasmuch as vanadium is widely distributed in soil and rock, some vanadium can be expected to be found in fresh water. In fresh water, the
vanadium is usually in solution as a salt, in the pentavalent state, in
which its compounds are most soluble. Therefore, in going from soil
or rock into water, the vanadium usually changes from a trivalent to
a pentavalent state.

Geographic differences in vanadium concentrations in surface water
or groundwater might be expected from effluents, leached by rainwater
from industrial plants or from natural vanadium sources, entering the

water table. Schroeder[156] indicates that the uranium ores of the Colorado Plateau supply its rivers with vanadium concentrations up to 70 μg/liter and that vanadium has been found in concentrations of 30–220 μg/liter in waters of Wyoming, some of which are used for drinking. Recent studies by Linstedt and Kruger[107] have shown that mean vanadium concentrations in natural waters range from 0.3 to over 20 μg/liter, as shown in Table 4-9. In addition, Strain *et al.*[185] report vanadium concentrations up to 19 μg/liter on the basis of spectrographic analysis of drinking-water samples obtained from nine New Mexico municipalities. T. S. Lovering (personal communication) has shown that the vanadium content of water from eastern Montana and the Dakotas varies from 7 to 150 μg/liter.

When the soluble vanadium in rivers reaches the sea, much of it is precipitated, and marine muds become enriched in vanadium. It has been estimated by Bowen[33] that only about 0.001% of the vanadium entering the oceans is retained in the soluble form in seawater.

PLANTS AND ANIMALS

The concentrations of vanadium in a variety of plants and animals are shown in Table 4-10. Generally, marine plants and invertebrate animals contain more vanadium than land plants, insects, and vertebrates. One

TABLE 4-8   Vanadium Concentrations in Various Rocks and Soils[a]

Source	Vanadium Concentration, mg/kg (ppm)
Rocks, by type	
Ultrabasic (peridotites)	200
Basic (shales)	200
Granite	40
Soils, by type	
Podzol (tundra)	100
Podzol	50
Forest soil	50
Clay	300
Soils, by country	
United States	200
U.S.S.R.	100
Great Britain	120
France	30
Japan	100
Earth's crust (average)	150

[a]Derived from Vinogradov.[202]

TABLE 4-9   Vanadium Concentrations in Natural Waters[a]

| Sampling Site | Vanadium Concentration, μg/liter | |
	Range	Mean
Sacramento River at Sacramento, Calif.	3.44– 4.31	3.93
Supply well in Fresno, Calif.	19.40–21.60	20.50
Green River at Flaming Gorge, Utah	0.20– 1.80	0.90
Colorado River at Page, Ariz.	1.50– 5.20	3.40
Colorado River at Hoover Dam, Nev.–Ariz.	1.80– 4.10	3.00
Colorado River at Parker Dam, Calif.–Ariz.	1.70– 4.30	3.00
San Joaquin River near Vernalis, Calif.	6.75– 7.01	6.87
Animas River at Cedar Hill, N.M.	0.20– 0.50	0.30
San Juan River at Shiprock, N.M.	0.50–49.20	7.50
Colorado River near Loma, Colo.	1.90–11.50	4.60
Colorado River at Yuma, Ariz.	1.00– 3.80	2.20

[a]Derived from Linstedt and Kruger.[107]

species of mushroom and one class of invertebrates, the ascidians (sea squirts), accumulate unusually large amounts of vanadium. The fly agaric mushroom, *Amanita muscaria*, has been shown by Bertrand[25] to contain about 100 times as much vanadium as other species of Amanita, other fungi, or higher plants studied. The vanadium concentration did not seem to be related to the vanadium concentration of the soil in

TABLE 4-10   Vanadium in Plants and Animals[a]

Source	Vanadium Concentration (dry wt), μg/g
Plants	
Plankton	5
Brown algae	2
Bryophytes	2.3
Ferns	0.13
Gymnosperms	0.69
Angiosperms	1.6
Bacteria	+
Fungi	0.67
Animals	
Coelenterates	2.3
Annelids	1.2
Mollusks	0.7
Echinoderms	1.9
Crustaceans	0.4
Insects	0.15
Fish	0.14
Mammals	< 0.4

[a]Derived from Schroeder.[158] (Data taken from Bowen.[33])

which the mushroom was growing, nor was the biologic role of vanadium elucidated. Of various higher plants, legumes seem to contain the most vanadium; the root nodules of legumes, which carry nitrogen-fixing bacteria, contain about three times as much as the rest of the plant.[25]

Prince[144] investigated the correlation between the vanadium of soil and vanadium concentration in corn grown in the soil, as shown in Table 4-11. There did not seem to be a clear relation between the two. However, Bertrand[25] has indicated that the vanadium content of plants is lowest in the aerial parts and higher in the roots, with the roots having approximately the same content as the soil in which the plant is growing.

Ascidians are very efficient accumu'.tors of vanadium.[142] The concentration of vanadium in these animals is 10,000 times that in the seawater in which they live.[80] In some ascidians, the vanadium is contained in green blood cells, called vanadocytes. In those cells, the vanadium is complexed to pyrrole rings, is associated with unusually high concentrations of sulfuric acid, and is in the trivalent state.[102] The function of this vanadium complex is not known; it does not seem to be involved in oxygen transport. In other ascidians, these vanadium complexes are found in blood plasma, rather than in specific cells.

Vanadium concentrations in the tissues of wild animals are shown in Table 4-12. Of those tested, the only tissue in which no vanadium was detected in any instance was the lung.

Human tissues generally accumulate less vanadium than those of wild

TABLE 4-11 Vanadium-Delivering Capacities of 10 New Jersey Soils[a]

Soil	Vanadium Concentration, μg/g	
	Soil	Corn
Norton	119	1.1
Annandale	90	0.83
Washington	65	0.37
Cossayuna	62	0.42
Croton	50	0.66
Coltz	46	0.36
Squires	40	0.40
Sassafras	33	0.49
Lansdale	20	0.70
Collington	11	0.76

[a]Derived from Prince.[144]

TABLE 4-12   Vanadium in Tissues of Wild Animals[a]

| Tissue[b] | No. Samples | Vanadium Concentration (wet wt), μg/g | |
		Mean	Range
Kidney	4	0.94	0.0–2.07
Liver	4	0.25	0.0–0.94
Heart	4	1.16	0.0–3.40
Spleen	1	1.16	–

[a]Derived from Schroeder.[158]
[b]From beaver, deer, woodchuck, rabbit, muskrat, and fox.

animals. Vanadium was found, however, in over half the lung samples studied, and an increase in vanadium was noted with increased age, especially in people over 60.[198] Next in frequency of occurrence of vanadium were the lower small and large intestines, omentum, and skin. No other tissue seemed to retain vanadium to any great extent. Although not listed in Table 4-13, no vanadium was detected in human aorta, brain, muscle, ovary, or testis.

There is significantly more vanadium in the lungs of people living in the Middle East than in those of people from the United States, Africa, and the Far East.[197] Furthermore, vanadium was found in 12 of 101 aortas and in 21 of 152 kidneys taken from subjects in the Middle East, whereas no vanadium was found in these tissues from American subjects.

FOOD

Vanadium is ubiquitous, so it is not surprising that it is contained in the foods that man consumes. What is surprising is the lack of information on the vanadium content of foods. Interest is lacking probably because vanadium, until recently,[86] was not considered to be essential to man or animals and, in the concentrations found in foods, has not caused toxicity problems.

Five references[13,24,120,161,177] were found giving information on the vanadium content of foods. The data represent the vanadium concentrations of foods from several countries. Table 4-14 compares data on food plants from the five reference sources. Although the marked differences could be due to the different concentrations of vanadium available in the soils, they are probably due to the differences in the analytic techniques used and the stage of development of the methods. The present method of choice for vanadium analysis at low concentrations is activation analysis.[99] The much higher values reported in three of the references were based on emission spectroscopy, whereas the lower and newer values reported by Söremark[177] were obtained by activation analysis. The wide discrepancy in vanadium concentrations is probably explained by systematic bias of the analytic techniques. The lower values are preferred, because they are less subject to errors caused by blanks and matrix effects, as discussed in Appendix B. Historically, as analytic procedures have been improved, concentrations of the element under investigation have had a tendency to be reduced. Although only one food value is given (pea), Mitchell[120] has reported later data indicating that lower values might be obtained with emission spectroscopy. These values would indicate that the concentrations of vanadium in plant foods are but a few parts per billion on a wet-weight basis, and not the approximately 1 ppm reported by other workers.

This low concentration is understandable if the uptake of vanadium-48 in vegetables is observed. Söremark[177] reported a range of uptakes

TABLE 4-13   Vanadium in Adult Human Tissues, United States[a]

Tissue	No. Samples	Samples Containing Vanadium		Vanadium Concentration (ash wt), µg/g	
		No.	%	Median	Maximum
Lung[b]	27	22	81.5	4	18
Lung	141	73	51.8	1	13
Cecum	31	14	45.2	< 1	4
Ileum	84	31	36.9	< 1	3
Rectum	42	11	26.2	< 1	2
Sigmoid colon	108	17	15.7	< 1	3
Omentum	75	11	14.7	< 1	3
Skin	21	3	14.3	< 1	< 1
Adrenal	13	1	7.7	< 1	< 1
Esophagus	66	5	7.6	< 1	< 1
Duodenum	67	5	7.5	< 1	< 1
Thyroid	21	1	4.8	< 1	< 1
Jejunum	102	4	3.9	< 1	< 1
Liver	148	5	3.4	1	< 1
Uterus	32	1	3.1	< 1	< 1
Bladder	110	3	2.7	< 1	< 1
Stomach	130	3	2.3	< 1	< 1
Prostate	50	1	2.0	< 1	< 1
Pancreas	139	2	1.4	< 1	< 1
Spleen	143	1	0.7	< 1	< 1
Fat	28	0	–	–	–
Heart	143	0	–	–	–
Kidney	122	0	–	–	–

[a]Derived from Schroeder.[158] (Data taken from Tipton and Cook.[195])
[b]From a study of vanadium in lung from people living in San Francisco.[198]

TABLE 4-14 Vanadium Concentrations in Food Plants

Plant	Vanadium Concentration (wet wt), ng/g				
	Ref 161	Ref 24	Ref 13	Ref 177	Ref 120
Lettuce	1,080	–	–	21.0	–
Radish	3,020	–	540	52.0	–
			400		
Apple	0	–	330	1.1	–
			210		
Tomato	0	–	–	0.027	–
Potato	1,490	–	70	0.82	–
			20		
Pear	50	–	24	< 0.1	–
			11		
Carrot	990	254	690	< 0.1	–
			1,120		
Beet	880	–	390	< 0.1	–
			650		
Pea	460	186	–	< 0.1	< 30[a]

[a]Dry-weight basis.

from $1.01 \times 10^{-5}\%$ of carrier free vanadium-48 in lettuce (dry wt) to $1.5 \times 10^{-3}\%$ in parsley. It appears that these plants did not take up much radioactive vanadium, and the inability of plants to take up larger amounts of soil vanadium probably accounts for the low concentrations observed in plant foods on analysis.

Söremark[177] also reported data on the vanadium concentrations of animal tissues, foods, milk, and gelatin. Seafoods, calf liver, and gelatin seemed to be relatively high in vanadium, 2.4–44 ppb (wet wt). Fresh trout, however, contained only 0.4 ppb. Fresh milk samples from six different locations contained approximately 0.1 ppb. Veal and pork from Stockholm were also low, containing less than 0.1 ppb (ash wt). Larger fish failed to show a measurable content of vanadium. This indicates that the vanadium concentration in the more commonly eaten foods might be rather low. The older literature contains higher values, but their accuracy is in question in view of the values obtained by the newer methods. Strain[184] has reported the results of Martinex, who, using colorimetric procedures, found vanadium concentrations of 20 μg/100 g in sardines and herring.

The data concerning the vanadium content of foods are very sparse indeed, but from these data we can draw the following conclusions: Vanadium is incorporated into foods from both plant and animal sources. The amount that a particular plant species takes up is variable and probably reflects to some extent the vanadium available in the soil.

The amount in animal foods probably reflects the diet. Seafoods are higher in vanadium, as are animal tissues in cases in which dietary vanadium has been increased.[86] Vanadium in foods is probably around a few parts per billion.

That we are interested in the total body burden of vanadium and the amount that foods contribute to this burden can be assumed. Obviously, more information, based on the best analytic procedures, is needed before this question can be answered intelligently: (1) The vanadium-uptake variability within one particular food-plant species must be known, as well as the influence of soil on availability; (2) whether any of the plants used as foods are vanadium accumulators must be determined; (3) the accumulation of vanadium in food plants due to foliar application of vanadium must be known and—more specifically—the uptake from the air or residue on the leaf surface of vanadium from polluted air; (4) the vanadium concentrations in foods derived from plants and animals are needed; and, finally, (5) whether polluted air and water significantly increase the vanadium content of animal tissues used as foods must be determined.

5

Biologic Effects of Vanadium

With the increasing production and use of vanadium, the handling of its different compounds has become an occupational-health problem throughout industrialized countries. Exposures have been observed in workers engaged in the manufacture of vanadium pentoxide from vanadium-concentrated slag, in the manufacture of ferrovanadium in pelletizing plants, in the cleaning of oil- and gas-turbine-fired boilers (both in plants and aboard ship), in refinery operations using vanadium as a catalyst, in glass and ceramic manufacturing, in electronics, and in smaller plant operations involving handling, grinding, and bagging.

Reviews on the toxicity of vanadium salts in man and animals have been published by Symanski,[188] Sjöberg,[173] Stokinger,[182,183] Faulkner Hudson,[60] and Athanassiadis.[10]

EFFECTS ON MAN

Industrial Exposure

Vanadium uptake in humans through ingestion or by direct injection into the circulatory system is uncommon, being encountered for the most part in experimental or investigative situations. The route of entry

46

most common in industrial exposures is the respiratory system, although entry can also be via the gastrointestinal system as part of the overall airborne exposure.

The first reliable description of the effects of vanadium dust in man was reported by Symanski.[188] He presented a critical review of the earlier literature and his own description of the symptoms based on 19 persons exposed for periods of a few months to several years. He concluded that there is a characteristic picture and that the symptoms are due entirely to the irritating effect of vanadium on the mucous membranes of the respiratory tract and on the conjunctiva. In all his cases, he observed conjunctivitis with smarting eyes, rhinitis with watery discharge, sore throat, persistent cough with pressure over the chest, possibly a stitch, and many rhonchi. He pointed out that severe chronic bronchitis might occur, possibly with bronchiectasis. He emphasized strongly that he had observed no gastrointestinal symptoms, no signs of lowered activity of the blood-forming organs, nor any symptoms of organic injuries of the kidneys or the nervous system. He believed that there was no reason to suspect an absorptive poisoning effect, but he did not present any tissue or body-fluid analyses or histologic data to support this contention.

Williams[207] reported an incident in which boiler cleaners sustained vanadium intoxication. The primary symptoms occurred 0.5–12 hr after starting work and consisted of rhinorrhea, sneezing, watering of the eyes, sore throat, and substernal soreness; the secondary symptoms appeared after 6–24 hr and consisted of a dry cough, wheezing, severe dyspnea, lassitude, and depression, with a disinclination to follow the usual evening activities. A greenish-black coating developed on the tongue; this faded 2–3 days after removal from contact with the petroleum soot. In some cases, the cough became paroxysmal and productive. The symptoms continued while the subjects were at work and did not become less severe until 3 days after removal from the exposure. In one man, the wheezing and dyspnea persisted for over a week, but no permanent effects were noted. The environmental data are shown in Tables 5-1 and 5-2. Table 5-1 indicates that there were 3,526 particles, up to 11 μm in diameter, per milliliter of air. Of these, 93.6% did not exceed 1 μm. The weight concentration of dust particles up to 1 μm was calculated to be 0.36 mg/m^3, or 2.9 wt % of particles not greater than 11 μm. This does not reflect the composition of the total dust, in which the presence of a large number of particles greater than 11 μm in diameter would greatly affect the weight distribution. Table 5-2 shows the analysis of the dust concentrations at various locations during boiler cleaning operations.

TABLE 5-1   Thermal-Precipitator Samples Taken from Superheater during Cleaning Operation[a]

Diameter of Particles, μm	No. Particles per Milliliter of Air	Numerical Concentration of Particles, %	Concentration of Particles	
			mg/m^3	wt %
0.15– 1.0	3,300	93.6	0.36	2.9
1.0 – 5.5	217	6.14	4.09	33.3
5.5 –11	9	0.26	7.85	63.8

[a]Derived from Williams.[207]

Browne[36] studied vanadium poisoning in 12 men exposed to exhausts from gas turbines using heavy residual fuels. The symptoms noted occurred between the first and fourteenth days of exposure and were generally similar to those described by Williams, except that there was bleeding, rather than running of the nose. Browne reported skin effects, but did not observe mental depression.

Browne and Steele[37] surveyed three catalytic oil-gas plants in which the nickel-impregnated catalyst became coated with fuel-oil ash containing vanadium. This coating had to be cleaned manually periodically. The catalyst had to be removed and sieved, and new catalyst had to be added to replace losses. The operation resulted in high concentrations of vanadium pentoxide dust. The men employed in the changing process were protected with goggles, respirators, and protective clothing.

Lewis[105] reported a study of 24 men who had worked with vanadium without exposure to other metals for more than 6 months. He was one of the first investigators to describe bronchospasm, which persisted in some cases for 2–3 days after cessation of exposure. As indicated by Tables 5-3 and 5-4, he found significant differences in symptoms and physical findings between controls and exposed workers. Eye, nose, and throat irritations, cough, wheezing, and productive sputum were among the most significant symptoms; injected pharynx,

TABLE 5-2   Analysis of Dust from Boiler during Cleaning[a]

Area	Dust Concentration, mg/m^3 of Air	Vanadium Concentration	
		% of Dust	mg/m^3 of Air
Superheater chamber	659	6.1	40.2
Superheater chamber	239	7.2	17.2
Combustion chamber	489	12.7	58.6

[a]Derived from Williams.[207]

green tongue, and wheezes, rales, or rhonchi were among the most significant of the physical findings. The effects of irritation of mucosal surfaces by vanadium were evident and consistent with the symptoms elicited. Wheezes, fine rales, and rhonchi were often heard on auscultation of the chest, and injection of the pharynx and nasal mucosa was common in the exposed group. He also noted a green coloration of the tongue. This benign "green tongue" in vanadium workers was first described by Wyers[209] in 1946. The harmless pigmentation, which is a result of the deposition of vanadium salts of reduced-valence forms, varied from a pale green to a dark greenish-black. Symanski[188] did not mention this finding. Sjöberg,[171] who had seen some of Wyers's subjects, reported that it was not observed in his group of workers. In Lewis's study,[105] it was observed in nine of 24 men. Table 5-5 summarizes the results of air sampling in Lewis's study. The average concentration of vanadium found in the urine of the control group was 11.6 μg/liter; the mean in the exposed group was 46.7 μg/liter.

Zenz, Bartlett, and Thiede[213] reported on 18 men engaged in pelletizing pure vanadium pentoxide. The clinical picture of *acute* illness appeared remarkably uniform among these workers. The syndrome consisted of a rapidly developing mild conjunctivitis, severe pharyngeal irritation, and nonproductive persistent cough, followed by diffuse rales and bronchospasm. After severe exposure, four of the men complained of skin "itch" and a sensation of "heat" in the face and forearms, but no objective evidence of skin irritation was found. It is possible that dermatitis did not develop among these workers because extremely cold weather prevented sweating and the intensity of the respiratory symptoms precluded prolonged exposure. A striking feature among the

TABLE 5-3 Symptoms in Vanadium Workers[a]

Symptom	Incidence, %		χ^2 Value
	Control	Exposed	
Cough	33.3	83.4	13.71[b]
Sputum	13.3	41.5	5.55[c]
Exertional dyspnea	24.4	12.5	0.592
Eye, nose, throat irritation	6.6	62.5	23.17[b]
Headache	20.0	12.5	0.124
Palpitations	11.1	20.8	0.538
Epistaxis	0	4.2	0.148
Wheezing	0	16.6	5.20[c]

[a]Derived from Lewis.[105]
[b]Significant beyond $p = 0.01$.
[c]Significant at $p = 0.02$.

TABLE 5-4 Physical Findings in Vanadium Workers[a]

Physical Finding	Incidence, %		x^2 Value
	Control	Exposed	
Tremors of hands	4.5	4.2	0.320
Hypertension	13.3	16.6	0.0002
Wheezes, rales, or rhonchi	0	20.8	6.93[b]
Hepatomegaly	8.9	12.5	0.003
Eye irritation	2.2	16.6	2.94
Injected pharynx	4.4	41.5	12.62[b]
Green tongue	0	37.5	14.53[b]

[a]Derived from Lewis.[105]
[b]Significant beyond p = 0.01.

acute intoxications was the increased severity of symptoms with repeated exposures of shorter duration and lower intensity. The severity
of the symptoms developing after second, third, or subsequent exposures was so dramatic as to suggest a sensitivity reaction, rather than
the effect of a cumulative dosage. Although these recurrent symptoms
appeared more severe than the original illness, there was no clinical
evidence that these exacerbations were of longer duration.

T. H. Milby (personal communication) described an incident in
which 21 workmen, most of them boilermakers, were assigned to a job
at a local refinery installing new catalytic-converter tubes. Each tube
contained a large number of pipes about 7/8 in. in diameter and 10 ft
long, which were to be filled with marble-sized pellets of vanadium
pentoxide. Each worker used a small measuring cup to drop the pellets
into the tubes in such a fashion as to prevent their clogging the channels
near the top. The workmen were supplied with dust masks and goggles.
The principal dust exposure seems to have occurred when the pellets
were dumped into a bin before the men dropped them into the con-
verter pipes. The job was carried on during three shifts. After about
72 hr, or 3 working days, of exposure, the workers began to complain
of nasal, ocular, and bronchial irritation. By the fourth day, most of
them were very ill. The most common symptoms were upper respiratory
irritation, eye irritation, chest pain, cough, and increasing shortness of
breath, which became severe enough to be disabling in every instance.
Other reported symptoms were blurred vision, ringing in the ears, diz-
ziness, palpitations, nausea, vomiting, deep bone pain, headache, and
loss of appetite. There were no permanent sequelae. Environmental
analyses indicated that the pellets contained 11.7% vanadium pentoxide
with a trace of cobalt. Dust samples taken during shaking of these pellets indicated that the average particle diameter was 1.1–1.5 μm. The
dust concentration during exposure was not given.

Sjöberg published a classic presentation[173] of his clinical and experimental investigation on the effect on man and animals of the inhalation of vanadium pentoxide dust. This was followed by reports[172,174] on chronic bronchitis, the possible risk of emphysema, and the effect of vanadium exposure on skin, eye, and respiratory tract. He presented a series of workers who had been exposed to vanadium dust during the manufacture of vanadium pentoxide and had been observed by him over a 2-year period. The symptoms consisted chiefly of slight conjunctivitis and irritation of the nasal mucosa—in most cases with nasal catarrh and moderate pathologic changes in the mucous membrane. Acute and chronic hyperplastic changes in the mucous membrane—in many cases of "allergic type"—predominated at the final examinations. The number of cases of objective changes in the mucous membrane did not appear to have increased during the observation period. Dryness and irritation of the throat appeared as the result of moderate changes in the mucous membrane. Chronic atrophic changes predominated at the final examination, and atrophic pharyngitis seemed to develop in one case during the observation period. The number of cases with significant pathologic changes was greater at the final examination than at the beginning of the observation period, at least in the "regulars" of the factory. The pathologic changes in the larynx and the trachea were very slight in most cases, as were the changes observed bronchoscopically. The most characteristic symptoms were coughing (in most cases violent, in some merely irritation of the throat) with or without expectoration, wheezing sounds in the chest (also heard on auscultation),

TABLE 5-5 Atmospheric Concentration of Vanadium in Various Plant Areas[a]

Location	Vanadium Concentration (as vanadium pentoxide), mg/m^3	Mass Respirable, %	Concentration of Particles < 5 μm, %
Fusion furnace No. 1, Colorado	0.097	51	99
Fusion furnace No. 2, Colorado	0.243	17	92.5
Precipitation, Colorado	0.204	16.6	96.5
Vanadium pentoxide drying, Colorado	0.105	19	97
Vanadium pentoxide dumping, Colorado	0.925	2	96.3
Foremen's office area, Ohio	0.018	100	100
Ammonium vanadate production bagging, Ohio	0.380	4	98.7

[a]Derived from Lewis.[105]

and dyspnea. Five cases of bronchopneumonia or pneumonia occurred among those who had worked constantly at the factory, and two cases among those who had left during the observation period. No chronic changes in the lungs (pneumoconiosis, fibrosis, emphysema) were demonstrated. There was no discoloration of the tongue, and there were no gastrointestinal symptoms. Increased blood and urine vanadium concentrations were demonstrated by spectral analysis. Palpitation of the heart on exertion and sometimes also at rest was present in several cases. A case of transient coronary insufficiency and the unexpectedly high incidence of extrasystoles were the remarkable features; it cannot be demonstrated that they were associated with the exposure to dust, but this possibility should not be ruled out. The blood pressure was not raised, nor was it increased during the observation period. Eczema was present and was in some cases probably attributable to hypersensitivity to vanadium (shown by patch tests). In one case, the eczematous reaction was verified by microscopic examination. The soda used in the factory may also have contributed to the development of the skin lesions, at least in the cases with negative patch-test results. Giddiness and neurasthenic symptoms were present in some cases; tremor was present in only one case. Sjöberg concluded, on the basis of thymol turbidity tests and the absence of liver lesions, that no causal relation with the exposure to vanadium could be justifiably postulated. No urinary tract symptoms could reasonably be associated with vanadium.

Tebrock and Machle[193] described "exposure of a plant population [plant workers] to a vanadium-bearing phosphor (europium yttrium orthovanadate) at a mean level of 0.84 mg/m^3 of air for five years [that] resulted in an increase of 30% [in the] annual rate of minimal injury by vanadium." Following are other results of the study:

- No chronic or systemic effects were found.
- No significant chest x-ray changes were seen in the 1960–1967 period.
- The incorporation of vanadium as the orthovanadate into the crystalline lattice of the phosphor apparently results in the formation of a compound of much lower toxicity than that of vanadium pentoxide alone.

Experimental Studies

There have been few experimental studies in which humans have been exposed to vanadium oxide dusts. Zenz and Berg[214] conducted such a study on nine healthy volunteer subjects, 27–44 years old. Before exposure, pulmonary-function tests were performed on three separate occasions to obtain baseline measurements. Determination of vanadium content in the biologic samples was made by the laboratory procedures described by Hulcher.[88]

In the first test, two volunteers were inadvertently exposed to vanadium pentoxide dust at 1 mg/m³, rather than the planned 0.05 mg/m³, and this exposure lasted for 8 hr. Some sporadic coughing developed after the fifth hour and at the time was believed to be psychologic. Near the end of the seventh hour, more frequent coughing developed in both subjects. By evening, persistent coughing had begun, and it lasted for 8 days. There were no other signs of irritation, nor was there fever or increased pulse rate. During the postexposure observation period, chest examinations revealed clear lung fields. Normal physical activities were pursued. Immediately after exposure and weekly for 3 weeks, lung-function tests were repeated; no differences from the baseline were detected. There were no alterations of the white-blood-cell counts or differential cell patterns. Urinalyses were normal. Nasal smears made 24 hr, 72 hr, and a week later did not disclose eosinophilia.

Three weeks after the original exposure, while preparing for another test, the same two volunteers accidentally experienced a 5-min exposure to a heavy cloud of vanadium pentoxide dust. Within 16 hr, marked coughing, with production of sputum, developed. On the following day, rales and expiratory wheezes were present throughout the entire lung field, but pulmonary functions were normal. Administration of a therapeutic dose of 1:2,000 isoproterenol for 5 min by positive-pressure inhalation relieved the coughing, and adventitious sounds were no longer present. Coughing resumed in about 1 hr and continued for a week. No other symptoms occurred during this period. and eosinophilia was absent from the blood and the nasal mucus.

On the basis of these unexpectedly severe clinical events, it was decided to set the planned exposure concentration during the next test at 0.25 mg/m³. Five volunteers were exposed to vanadium pentoxide at 0.2 mg/m³ ± 0.06 SD for 8 hr. Microscopic particle size analysis of the dust indicated that more than 98% of the particles were smaller than 5 μm in diameter. Even at this concentration, all men had developed a loose cough by the following morning. Physical examinations were unrevealing, and there were no other systemic complaints. All subjects continued normal activities both at home and at work. Daily observations uncovered no abnormalities. Several subjects stopped coughing after a week, and all had ceased coughing by the tenth day. No changes were noted between pulmonary-function studies conducted immediately before and after the test period. Spirometry was repeated at the end of 2 weeks, without detectable changes from pre-exposure studies. Differential white-cell counts were normal. Blood samples for vanadium and other components were drawn every other day for 8 days, and daily urine samples were collected. After a week, blood and urine samples were taken every other day for 14 days. No vanadium was detected in the blood of any subject. The greatest amount of vanadium, 0.013 mg/100 ml, was found in the urine 3 days after exposure; none was detectable a week after exposure. Maximal fecal vanadium was 0.003 mg/g; none was detected after 2 weeks.

Because of the unexpected responses in men at 0.2 mg/m³, two volunteers not previously exposed to vanadium pentoxide were exposed for 8 hr at a concentration of 0.1 mg/m³. At this concentration, no symptoms occurred during or immediately after exposure. Within 24 hr, however, considerable mucus formed. This mucus was easily cleared by slight coughing, which increased after 48 hr, subsided within 72 hr, and disappeared completely after 4 days. There were no deviations in laboratory studies. The acute manifestations of marked pulmonary irritation observed in all subjects demonstrated the prompt reaction of humans to fine vanadium pentoxide dust. Despite the distinct clinical picture of pulmonary irritation displayed by these subjects, the pulmonary-function tests remained unaltered. No systemic aberrations were observed. When two of the volunteers were re-exposed for 8 hr at a concentration of 0.1 mg/m³, a distinct clinical picture of pulmonary irritation appeared.

Industrial operations involving welding procedures and metal pouring are expanding at an enormous rate throughout the world. Steel and Sanderson[180] studied production of toxic substances during standard arc welding and demonstrated that copper, lead, vanadium, chromium, and zinc were present in coatings of the 12 electrodes studied. Little information exists on toxic substances present as impurities, and even less on their concentrations in arc-welding fumes. The frequent occurrence of multiple exposures to fumes of metal oxides was pointed out by Fishburn and Zenz.[61] Laboratory studies on a welder who had received heavy exposure to zinc oxide fumes indicated normal serum glutamic oxalacetic transaminase (SGOT) and serum glutamic pyruvic transaminase (SGPT), but a zymogram (lactic acid dehydrogenase) indicated that fraction 3 (pulmonary fraction) was increased. This altered zymogram pattern was interpreted as being due to a chemical exposure.

Excretion

Vanadium is excreted mainly through the urine and to some extent in the stool.[55] Jaraczewska and Jakubowski[91] attempted to correlate degree of exposure to vanadium pentoxide with the concentration of vanadium excreted in the urine. A summary of their data is shown in Table 5-6. They concluded that the concentrations in urine and in the air could not be correlated with sufficient certainty.

Watanabe et al.[204] studied 13 workers who had been exposed to vanadium pentoxide at 0.48–2.65 mg/m³ for 1–3 years. Table 5-7 shows that the exposure may have had an effect on ascorbic acid metabolism.

In spectrographic analysis of organs and muscle tissues of Americans, vanadium has been detected only in lung and intestine; most positive samples showed 0.01 μg or less of vanadium per gram of tissue. Modern and more sensitive methods of detection would probably reveal vanadium in other tissues as well, but at low concentrations. Vanadium appears to be present mostly in fat and bone, but measurable amounts occur in serum, according to Schroeder et al.,[161] and Valberg and Holt[201] detected it in human red blood cells with neutron-activation analysis.

Differential Diagnosis

With regard to diagnosis, it is not always easy to distinguish the disease picture from acute infection of the respiratory tract. The history of exposure to vanadium dust must be ascertained. Determination of the content of vanadium in the blood and urine, the cystine content of fingernails, and other biochemical indices, as discussed later, should be taken into consideration when making a diagnosis. If eczema is present, it should be considered to be caused by vanadium only after patch tests with positive results. Stokinger[183] pointed out that vanadium at extremely low tissue concentrations (1 μg/g of tissue) can induce derangements of basic metabolic processes that are not manifest clinically or felt by the subject. Stokinger and his colleagues developed a highly sensitive test of pretoxicosis that proved to be of practical value in the management of the health of vanadium workers. This test was based on the fact that the cystine content of the fingernails is decreased after chronic low-grade exposure to vanadium. Using a sensitive technique, they demonstrated cystine decrease even when the amount of urinary vanadium resulting from exposure amounted to no more than about 20 or 30 μg/liter.

Effect of Form of Vanadium on Toxicity

Roshchin[154] pointed out that the degree of vanadium toxicity depends on the dispersion and solubility of vanadium aerosols in biologic media. He also noted that the toxicity depends on valence; i.e., it increases

TABLE 5-6 Concentrations of Vanadium in Air and in Urine of Workers in Three Chemical Plants[a]

| Plant | Vanadium Concentration in Air, mg/m³ | | | | Measured through Dust Mask | | | | Vanadium Concentration in Urine, μg/liter | | | |
| | Measured Directly | | | | | | | | | | | |
	No. Determinations	Min.[b]	Max.	Mean	No. Determinations	Min.[b]	Max.	Mean	No. Determinations	Min.[b]	Max.	Mean
A	4	5	117	63	1	—	—	0.9	6	84	500	230
B	8	0	5	2.3	4	0	0.1	0.025	5 / 6[c] {	0 / 0	50 / 265	10 / 90
C	11	0	0.7	0.3	4	0	0.03	0.007	6 / 4[c] {	0 / 0	70 / 100	17 / 42

[a]Derived from Jaraczewska and Jakubowski.[91]
[b]Zero represents a negative result—i.e., below 0.025 mg/m³ in the air or below 50 μg/liter in urine.
[c]Urine specimens collected at the beginning of the working day (16 hr after the end of exposure); the remaining urine samples were collected immediately at the end of work.

56

TABLE 5-7   Vanadium and Ascorbic Acid Excretion in Workers
Exposed to Vanadium Pentoxide Fumes[a]

Urinary Excretion	Maximum	Average	Minimum
Vanadium, μg/liter			
Exposed workers	259.3	92.7	21.0
Controls	11.0	6.9	3.5
Increase over controls, factor	23.6	13.4	6.0
Ascorbic acid, μg/3 hr			
Exposed workers	4,000	1,900	500
Controls	9,700	2,500	700
Decrease from controls, %	58.8	24.0	28.6

[a]Derived from Watanabe et al.[204]

with increasing valence, with pentavalent vanadium being most toxic.
In addition, it is toxic both as a cation and as an anion.

Community Exposure

In a statistical study by Hickey et al.,[82] concentrations of vanadium and
nine other metals in the environmental air of 25 communities in the
United States were correlated with mortality indices (1962–1963) of
eight categories of prevailing diseases. Various techniques of correlation
were used, including canonic analysis. From limited data, it appeared
that vanadium—as well as cadmium, zinc, tin, and nickel—correlated
fairly well with the several disease incidences studied, e.g., "diseases of
the heart," nephritis, and "arteriosclerotic heart." Moreover, tests of
statistical significance of various combinations showed that the addition of vanadium to cadmium produced a reduction of more than 10%
(the greatest reduction) in the error of variance. There was a very high
intercorrelation between vanadium and nickel, which was not explained
in the study.

In another statistical study, by Stocks,[181] mortality from lung cancer
was correlated with the concentration of particles of 13 trace elements
in 23 localities in Great Britain. The findings of the study were as follows: Vanadium, arsenic, and zinc showed weak associations with lung
cancer. Vanadium showed a strong association (second only to those of
beryllium and arsenic) with bronchitis in males. Vanadium and beryllium were found to be associated with pneumonia. Vanadium, beryllium, and molybdenum showed correlations with other cancers, but
only in males. Thus, vanadium ambient-air concentrations showed statistical correlations with bronchitis, pneumonia, lung cancer, and other

cancers. The estimated vanadium correlation coefficients were as follows: bronchitis, males, 0.620; pneumonia, males, 0.805 (highest among 11 metals); pneumonia, females, 0.711 (highest except for beryllium); lung cancer, males, 0.770 (highest except for beryllium); and other cancers, except of stomach, males, 0.556 (highest).

Both studies mentioned above represent efforts to test statistically for significant correlations between environmental concentrations of a number of trace elements and mortality rates related to various diseases in urban centers. Hickey et al.[82] used more sophisticated statistical techniques. Their study used canonic rank correlation to test for the combination of variables that would maximize the correlation between pollutants and diseases; it also tested for and found significant intercorrelations, thus reducing the validity of conclusions based on the estimated numerical values of the correlation coefficients. However, the study by Stocks considered diseases that are better defined than most of those considered in the study by Hickey et al. Furthermore, it considered such important characteristics as population density, sex, and age.

The two studies cannot be compared, because the correlations were determined for different diseases (except for lung cancer), and the sets of pollutants considered were not identical. But it is interesting to note the differences in the correlation coefficients estimated by the two studies with respect to the relation between lung-cancer mortality and vanadium concentrations: 0.320 by Hickey et al.[82] and 0.770 by Stocks.[181] Hickey found that, when vanadium was considered with cadmium, the multiple correlation coefficient was 0.767. He points out that, in addition to cadmium, vanadium in the ambient air may contribute to diseases of the heart. Both studies were handicapped by the lack of long-term data, the omissions of important pollutants (e.g., organic compounds) that are known to be causally related to some of the diseases considered and that cannot be assumed to be constant, the unsatisfactory definitions of disease, and the known and unknown intercorrelations (whether positive or negative) among the pollutants and the diseases considered.

During the early 1960's, large numbers of United States military personnel and their families stationed in Japan developed a high incidence of upper respiratory symptoms—a condition referred to as "Yokohama asthma." Many factors were investigated in the search for the etiologic agent, and vanadium was one of the agents studied. Vanadium was suspected because of the new sources of Japan's oil supply after World War II. Before the war, fuel oil low in vanadium had been obtained from Indonesia. This source dwindled, and postwar arrangements were made with Middle East suppliers whose oil contained considerably more

vanadium (see Chapter 3). Vanadium was never proved to be the etiologic agent accountable for "Yokohama asthma," but it may have been
a contributing factor.

Threshold Limit Values

In 1961, the American Conference of Governmental Industrial Hygienists (ACGIH) listed the following limits for vanadium.

	mg/m^3
vanadium pentoxide (dust)	0.5
vanadium pentoxide (fume)	0.1
ferrovanadium (dust)	1.0

On the basis of new data, intended changes were noted in 1970, and the
following limits were set in 1971.[6]

	mg/m^3
vanadium pentoxide (dust)	0.5 (ceiling)
vanadium pentoxide (fume)	0.05

Threshold limits are based on the best available information from industrial experience and from experimental human and animal studies.
The ACGIH Committee on Threshold Limit Values has documented the
pertinent scientific information that was used to base each limit.[6] The
Russian literature proposes more stringent limits—e.g., a reduction down
to 0.1 mg/m^3 for vanadium dust.[139] Their documentation stems principally from animal studies, most often rabbits. Schumann-Vogt[162] recommended that the maximal acceptable concentration for vanadium
pentoxide dust be set at 0.5 mg/m^3; for fumes, 0.1 mg/m^3; and for
ferrovanadium dust, 1 mg/m^3.

Stokinger[183] reported an animal study designed to test the suitability
of the threshold limit of 0.5 mg of vanadium pentoxide dust per cubic
meter of air for industrial conditions recommended by the Russians.
It was found that dogs, rats, guinea pigs, and rabbits tolerated such
exposure for 6 months at 6 hr/day without evidence of histologic
change referable to inhalation of the dust.

EFFECTS ON EXPERIMENTAL ANIMALS

Toxicity

The toxicity of vanadium has been found to be high when it is given
parenterally, low by mouth, and intermediate by the respiratory tract.

Roshchin et al.,[155] using albino mice, established the oral LD_{50} of vanadium trioxide (vanadium sesquioxide, V_2O_3) at 130 mg/kg, and that for vanadium pentoxide and vanadium trichloride at 23 mg/kg. Faulkner Hudson[60] summarized the toxicity of vanadium pentoxide when given parenterally. Table 5-8 shows the lethal doses of various vanadium compounds given by injection to four different species of laboratory animals—intravenously to rabbits, subcutaneously to the other animals. Faulkner Hudson noted that the rat and mouse are relatively resistant, whereas the rabbit is especially sensitive to vanadium compounds.

Mitchell[121] studied the influence of pH on the toxicity of vanadium in mice. It appears that the detoxication of vanadium can be affected by the bicarbonate reserve of the blood. Excessive blood bicarbonate may retard the process of converting vanadium to a less toxic state. Acidification of vanadium solutions before injection markedly reduced their toxicity.

Stokinger[183] reported that skin absorption occurs from approximately saturated (20%) solution of sodium metavanadate; its application to rabbit skin caused irritation. Although the skin does come into contact with vanadium particles in the environment, this route of absorption appears to be of minor importance.

Faulkner Hudson[60] reviewed earlier literature on the inhalation toxicity of vanadium pentoxide in which it was reported that cats tolerated exposure to 43 mg/m³ for at least 3 hr, but succumbed after 23 min at 500 mg/m³. Gastroenteritis and pneumonia, sometimes with edema of the lung, were seen at autopsy. In similar experiments with rabbits, inflammatory changes in the respiratory, digestive, and urinary tracts were found; vanadium appeared almost at once in the urine, indicating its rapid absorption into the circulation. Rabbits exposed to very pure

TABLE 5-8 Lethal Doses of Vanadium Compounds[a]

Compound	Lethal Dose, mg/kg			
	Rabbit	Guinea Pig	Rat	Mouse
Colloidal vanadium pentoxide	1–2	20–28		87.5–117.5
Ammonium metavanadate	1.5–2.0	1–2	20–30	25–30
Sodium orthovanadate	2–3	1–2	50–60	50–100
Sodium pyrovanadate	3–4	1–2	40–50	50–100
Sodium tetravanadate	6–8	18–20	30–40	25–50
Sodium hexavanadate	30–40	40–50	40–50	100–150
Vanadyl sulfate	18–20	35–45	158–190	125–150
Sodium vanadate		30–40	10–20	100–150

[a]Derived from Faulkner Hudson.[60]

vanadium pentoxide dust at 205 mg/m³, with nearly all the particles below 10 μm in diameter, died in 7 hr. Tracheitis was very marked, and there was bronchopneumonia with pulmonary edema. In addition, conjunctivitis, enteritis, and fatty degeneration in the liver were noted. Vanadium dust particles in small quantity were present in the lungs, and vanadium was detected in the ashed lung, liver, kidney, and intestine. Sjöberg[173] conducted long-term repeated studies in which rabbits were exposed 1 hr/day to vanadium pentoxide at 20–40 mg/m³ for several months. The pathologic changes noted included chronic rhinitis and tracheitis, emphysema, and patches of lung atelectasis with bronchopneumonia. In some cases, pyelonephritis was seen. Vanadium could again be detected in the lung, liver, and kidney, but not in the intestine, as in the heavily exposed group. Continuous exposure to a concentration of 10–30 mg/m³ was distinctly toxic to rabbits, causing bronchitis and pneumonia, loss of weight, and bloody diarrhea. With rats, 10 mg/m³ was toxic, and a small exposure of 3–5 mg/m³ caused symptoms after 2 months; a lethal exposure was considered to be 70 mg/m³ if prolonged for more than 20 hr. Injection of vanadium pentoxide into the trachea of rats caused pulmonary edema and purulent pneumonitis; the adrenals became congested and showed a reduction in lipid content.

In a series of experiments, Roshchin[154] and Roshchin et al.[155] exposed rabbits to vanadium trioxide aerosol at 40–75 mg/m³ for 2 hr/day over a period of 9–12 months. The animals showed signs of a general toxic effect, such as profuse nasal discharge of mucus and sneezing; attacks of bronchial asthma were noted in some animals. After 1.5 months, the animals exhibited dyspnea and tachypnea. At the termination of the experiment, the rabbits showed an average weight loss of 4.6% (the controls had gained 12.3%), a progressive decline in the number of white cells, a reduction in hemoglobin from 75% to 67.8%, a significant reduction in vitamin C content of the blood, a reduction of serum protein sulfhydryl groups, and a substantial reduction of tissue respiration in liver and brain. The cholinesterase activity increased by an average of 25%. No changes were noted in total protein, ratio of albumin to globulin or respiratory quotient. Rabbits were also exposed to vanadium trioxide at 40–70 mg/m³, vanadium pentoxide at 8–18 mg/m³, and vanadium carbide at 40–80 mg/m³ for 2 hr/day for 9–12 months. Vanadium pentoxide was 3–5 times as toxic as vanadium trioxide, but the acute and chronic intoxication that developed was similar to that reported for vanadium trioxide in its biochemical, functional, and morphologic abnormalities.

Chronic poisoning with the dusts of the compounds studied resulted in blood-protein changes: There was a decrease in albumins and an in-

crease in globulins, so that the albumin : globulin ratio was halved. By
the end of the eleventh month, additional effects included an increase
in the serum content of amino acids (cysteine, arginine, histidine), a
10% increase in the nucleic acid in the blood, a 29.8% decrease in the
serum content of sulfhydryl groups, a 50% decrease in the blood content of vitamin C, a considerable increase in the blood content of
chloride, and a drastic inhibition of tissue respiration in the liver and
brain.

Rats exposed to vanadium trioxide showed changes in the respiratory
system that included suppurative bronchitis, septic bronchopneumonia,
pulmonary emphysema, formation of cellular dust foci with signs of
necrobiosis in the phagocytes, and moderate interstitial pulmonary
sclerosis.

Rabbits and rats exposed to vanadium chloride, VCl_3, showed more
marked histopathologic changes in internal organs than those exposed
to vanadium trioxide. They were characterized by protein and fatty
dystrophies of the cells of liver, kidney, and myocardium; partial ne-
crosis of the tissues of some organs; and reduction in the ribonucleic
acid and deoxyribonucleic acid content of the cells of liver, kidney,
myocardium, stomach, intestine, and lung. Thus, vanadium chloride
proved to be more toxic than vanadium trioxide under the same experimental conditions.

The aerosols of vanadium, vanadium carbide, and ferrovanadium
were not considered highly toxic. However, they produced some local
and general physiologic reactions, such as catarrhal bronchitis, pathologic tissue proliferation, a moderate degree of interstitial pneumo-
sclerosis, marked catarrhal gastritis (a local effect occurring after oral
administration), and histopathologic alterations in the parenchyma,
such as local nephritis, fatty dystrophy of the hepatic cells, sclerosis
of hepatic and renal interstitial tissues, and perivascular edema in the
myocardium (a typical general toxic effect). Exposure to ferrovanadium dust was found to cause slight (statistically insignificant) changes
in the blood, such as a decrease in the sulfhydryl groups and nucleic
acids, but marked and significant proteinuria. The vanadium of ferrovanadium was found to be considerably more toxic than free vanadium because of its higher solubility in biologic material. Exposure
to vanadium carbide and vanadium dust was found to produce slight,
unstable, statistically insignificant biochemical changes in the blood.[154]

In all cases, the chief damage was to the respiratory tract, but no
fibrotic changes or specific chronic lesions were observed in the lungs.
The presence of vanadium in several organs was evidence of its absorption; this absorption was probably rapid, inasmuch as the lungs in the

long-term experiments contained no more particles than did those of animals exposed for a short period. Symptoms of vanadium toxicity consisted of depression of the respiratory center, constriction of the peripheral arteries of the viscera, hyperperistalsis, and enteritis. Other effects included congestion with focal hemorrhages in the lungs, fatty degeneration of the liver, and damage to the convoluted tubules of the kidneys.[60]

Studies on the effect of vanadium compounds when added to the diet of chicks and rats have been conducted by Romoser *et al.,*[153] Nelson *et al.,*[134] and Berg.[18] It was found that vanadium added to a corn–soybean meal diet at 20 ppm was tolerated by the chick, whereas 30 ppm or more resulted in significant growth depression. Berg *et al.*[19] reported a slight loss in body weight and a change in egg albumen quality in laying hens when they consumed a natural diet containing vanadium at 10 ppm added as ammonium metavanadate. Mortalities occurred when 200 ppm or more was added. Berg[17] pointed out that the composition of the diet influenced the severity of toxic symptoms when vanadium compounds were added, toxicity being greater when a purified diet was fed.

Mitchell and Floyd[122] tested ascorbic acid and ethylenediaminetetra-acetate as antidotes in experimental poisoning. In mice, rats, and dogs, both agents were effective antidotes, with ascorbic acid acting more rapidly.

Lillie[106] has reviewed the literature on vanadium toxicity as an air pollutant affecting the performance of *domestic* animals. Only one field case was mentioned. Heege[78] reported signs of vanadium toxicity in cows grazing in an area exposed to fuel-oil soot from a nearby factory. The soot, which came from cleaning oil-fired boilers, had been dumped into a ditch that ran through the meadow where the cows grazed. Dry weather and strong winds combined to spread a film of powdery soot over the meadow for dozens of meters down the ditch.

Tolerance

Daniel and Lillie[51] demonstrated that some tolerance to vanadium is developed by animals; increasing doses were tolerated in amounts far greater than were lethal on the first or second dose.

Essentiality

Hopkins and Mohr[86] reported data indicating that vanadium is an essential element for chicks. When chicks were maintained in a clean, all-plastic environment and fed a vanadium-deficient diet (10 ppb), they observed reduced feather growth and lowered blood cholesterol. Strasia[186] and Schwarz and Milne[163] substantiated this finding in rats and have reported a significant reduction in the rate of growth due to vanadium deficiency.

Hopkins and Tilton[87] showed that animals were able to metabolize low concentrations of vanadium easily, as would be expected of an essential element. They measured the distribution of trace amounts of intravenously injected vanadium-48 in selected organs and in subcellular liver particles in rats. Liver, kidney, spleen, and testes accumulated vanadium-48 for up to 4 hr and retained most of the radioactivity for up to 96 hr, at which time most other major organs retained only 14–84% of their 10-min uptake. After 96 hr, 46% of the vanadium-48 had been excreted in the urine and 9% in the feces.

Radioactivity in the liver subcellular supernatant fraction decreased from 57% to 11% of the liver radioactivity by 96 hr, whereas the mitochondrial and nuclear fractions increased from approximately 14% to 40%. The microsomal fraction changed little. Marked liver retention of vanadium-48 was probably due to its migration into the mitochondrial and nuclear fraction. No major differences were seen in either the rate or the amount of uptake of the three oxidation states of vanadium injected.

Söremark et al.[179] reported that, when radiovanadium was injected subcutaneously into young rats, the highest uptake was found in areas of rapid mineralization in the dentin and bone. Using adult mice, they noted a concentration of radioactive vanadium in the fetus, as well as in the bones and teeth. Söremark and Üllberg[178] noted that the mammary glands showed a high uptake, as did the liver, renal cortex, and lung. Underwood[199] also noted that some ascidians and tunicates may contain variable but usually high concentrations of vanadium. Vanadium may be physiologically functional to the animal, but its exact role is still unknown.[43]

Schroeder et al.,[161] using emission spectrographic analyses, reported the vanadium content of a number of animal tissues. They found values ranging from nondetectable amounts in some tissue up to 3.4 μg/g (wet wt) in domestic rabbit heart. They stated that animal fats have an affinity for vanadium, although Hopkins and Tilton,[87] using the radioisotope-uptake technique, were unable to confirm this finding. These values are probably high in light of Söremark's work.[177] Using activation analysis, he found that vanadium content in animal tissues ranged from less than 1 ppb (wet wt) to a few parts per billion. Again, using activation analysis, Hopkins and Mohr[86] reported that liver, kidney, and heart tissues from chicks on a deficient diet contained vanadium at only a few parts per billion, whereas the supplemented controls had 10 times as much.

The above information indicates that vanadium is an essential element for the chick and rat; and it is probably essential for other animals. On the basis of limited data, it appears that animal tissues contain vanadium at only a few parts per billion.

EFFECTS ON METABOLISM

Cystine and Cysteine Metabolism

Once vanadium enters the body, it affects cystine and a number of different metabolic reactions. Mountain et al.[127,128] showed that rats fed vanadium pentoxide at dietary vanadium concentrations of 25–1,000 ppm had a lowered cystine content in their hair, and workers exposed to vanadium, a lowered cystine content in their fingernails. These findings lead to the conclusion that the synthesis of cystine (and its reduction product cysteine) is decreased in animals and humans exposed to vanadium in concentrations higher than those usually found under normal environmental conditions. Bergel et al.[20] have shown that the destruction of cystine and cysteine is increased in the presence of vanadium. Metabolic processes that depend on cystine or cysteine should be decreased under conditions of increased vanadium intake.

Coenzyme A Synthesis

Mascitelli-Coriandoli and Citterio[113,114] showed that the coenzyme A content of rat liver was appreciably reduced by administration of sodium metavanadate either in the diet or by intraperitoneal injection. One of the compounds involved in the synthesis of coenzyme A is thioethanolamine, which is derived from cystine by decarboxylation. Therefore, it can be concluded that a reduction in the synthesis of cysteine by vanadium decreased the synthesis of coenzyme A.

Cholesterol Synthesis

Coenzyme A is involved in a variety of metabolic processes that use acetate as a starting product; it is possible that these processes will be diminished when excessive vanadium is administered. Curran[46] showed that vanadium decreased the synthesis of cholesterol from [^{14}C] acetate in rat liver. In 1961, Azarnoff et al.[11,12] showed that vanadyl sulfate,

$VOSO_4$, inhibits the enzyme squalene synthetase, which catalyzes the conversion of farnesyl pyrophosphate to squalene. Squalene is an intermediate in the conversion of acetyl coenzyme A to cholesterol. Vanadium also inhibits cholesterol biosynthesis *in vivo*.[46] Curran and Costello[49] showed that, in rabbits made atherosclerotic by a high-cholesterol diet, vanadium induced a more rapid mobilization of aortic cholesterol than was the case in control animals. Studies with humans also indicated that vanadium could lower body cholesterol.[29] Curran *et al.*[47] showed that, when a group of young men were given diammonium oxytartratovanadate over a 6-week period, a statistically significant lowering of serum total and free cholesterol concentrations occurred. In addition, Lewis[103, 105] described a lowering of serum cholesterol content in men exposed to dust high in vanadium.

However, in a controlled experiment by Somerville and Davies,[176] 12 patients with hypercholesterolemia and ischemic heart disease treated with oral ammonium vanadotartrate for 6 months did not show any change in serum cholesterol content or lipoprotein content, although toxic side effects of the administered vanadium occurred in six patients. Schroeder *et al.*[161] also showed the same lack of effect of vanadium on serum cholesterol content in a group of patients over 60. Curran and Burch[48] then turned to studies in older rats and found that, by using the livers of rats weighing 400–600 g, they could demonstrate inhibition of cholesterol synthesis by vanadium *in vitro* but not *in vivo*.

Recently, Curran and Burch[48] indicated that mitochondrial acetoacetyl coenzyme A deacylase, an enzyme that depresses the conversion of acetate to cholesterol, was activated by vanadium in young animals, but was inhibited by vanadium in older animals. These results may explain the contradictory findings on the effects of vanadium on cholesterol synthesis in man and animals.

Triglyceride and Phospholipid Synthesis

Acetyl coenzyme A is a precursor of fatty acids; therefore, vanadium might affect the synthesis of triglycerides and phospholipids. Triglycerides were found to be decreased in the livers of rats receiving vanadium,[46] but their serum concentration increased during vanadium-feeding experiments in man.[47] Snyder and Cornatzer[175] noted that intraperitoneal injection of vanadyl sulfate into rats immediately before injection of $H^{32}PO_4^{-2}$ resulted in decreased incorporation of phosphate into liver phospholipids within 3 hr. They speculate that the decreased incorporation resulted from inhibition of synthesis of phospholipids by vanadium and perhaps also from augmentation by vanadium of oxidative degradation of phospholipids, as suggested by Bernheim and Bernheim.[21,22]

Coenzyme Q Synthesis

A process requiring coenzyme A is the synthesis of coenzyme Q (ubiquinone), a benzoquinone with a polyisoprene side chain. Coenzyme Q is involved in the electron-transport system of mitochondria. Aiyar and Sreenivasan[4] have shown that administration of vanadium did, in fact, reduce the quantity of coenzyme Q in mitochondria. When cysteine was given with vanadium, the effect of the metal on coenzyme Q synthesis was reversed to some extent, and the simultaneous administration of adenosine triphosphate (ATP), coenzyme A, and cysteine with vanadium completely counteracted the metal's depression of coenzyme Q synthesis.

Many other biologic compounds require coenzyme A for their synthesis, but the effects of vanadium on these processes have not been studied.

Dental Caries

A study by Geyer[66] indicated that the addition of small amounts of vanadium to the diet decreased the incidence of dental caries in hamsters. However, this result could not be corroborated by Hein and Wisotzky[79] or, in rats, by Muhler.[130] Tank and Storvick[191] indicated that there was a relation between vanadium concentration in drinking water and the incidence of caries in children in Wyoming. Recently, McLundie et al.[116] reported that, of a variety of metal ions tested, only vanadium decreased the solubility of tooth enamel in an acid buffer in vitro. Hadjimarkos[72,73] concludes that the evidence is still too meager to confirm a positive relation between vanadium intake and the incidence of dental caries in humans.

Monoamine Oxidase

There is evidence that vanadium salts are activators of or coenzymes for tissue monoamine oxidase. Perry et al.[140,141] report that vanadium (10^{-5} M) accelerated by 125% the oxidation of tryptamine in the presence of crude preparations of monoamine oxidase obtained from fresh guinea pig liver and kidney. In comparison, 10^{-3} M ferrous and cobaltous ions increased activity by 20% and 33%, respectively. Inasmuch as serotonin (5-hydroxytryptamine) is also oxidized by mono-

amine oxidase, Schroeder[159, 160] has noted the possible importance of
vanadium salts in hypertension in man and animals. Schroeder found
that the urinary excretion of vanadium was 3 times greater in hyper-
tensive patients than in normal subjects.

However, Lewis[104] found that injection of sodium metavanadate at
2.5–5.0 mg/kg into dogs resulted in a decrease in urinary excretion of
5-hydroxyindoleacetic acid, the oxidation product of serotonin. This
dosage of vanadium was toxic to the dogs, and it is therefore possible
that the vanadium at this concentration inhibited monoamine oxidase,
rather than activating the enzyme, as it does at lower concentration.
In addition, the *in vitro* oxidation of serotonin and related amines was
catalyzed by vanadium in the absence of an enzyme.[112]

Hemopoiesis

Data on the effect of vanadium on hemoglobin content and anemia have
been somewhat conflicting. When vanadium compounds were administered to animals or humans that had a normal hemoglobin content,
there was little effect on hemoglobin concentration.[62, 103] But Myers
and Beard[131] showed that vanadium and several other trace minerals
were beneficial in treating nutritional anemia in rats. Administration of
iron alone to nutritionally anemic rats resulted in the return of hemoglobin content to normal in 6 weeks. But when the diet was supplemented with iron and a small quantity of vanadium, hemoglobin
concentration returned to normal in 2–3 weeks. In further related
experiments, Beard *et al.*[14] indicated that red cells seem to mature
more quickly in the presence of vanadium than in its absence. Strasia[186]
has reported that rats fed a diet containing less than 100 ppb had significantly increased packed-cell volume, blood iron, and bone iron,
compared with groups receiving 0.5, 2.5, and 5.0 ppm supplementally.
Further work seems necessary to explain how vanadium affects iron
metabolism.

Other Metabolic Effects

Vanadium has been shown by Hathcock *et al.*[76, 77] to uncouple oxidative
phosphorylation. Addition of ammonium metavanadate to give 25 ppm
in the diet of chicks led to uncoupling of oxidative phosphorylation in
isolated liver mitochondria. Uncoupling phosphorylation would result
in less A TP production, and these investigators speculate that vanadium
toxicity might be related to the uncoupling. In addition, Aiyar and
Sreenivasan[4] showed that vanadium salts reduce succinoxidase activity,

which would also result in reduced A TP synthesis. It should be noted that succinoxidase requires sulfhydryl groups for its activity.

Finally, Mascitelli-Coriandoli and Citterio[114] showed that vanadium could reduce the level of thioctic acid in rat liver, particularly in the mitochondrial fraction, where thioctic acid was reduced by 86% below that in control animals receiving no vanadium.

SUMMARY

The major signs and symptoms of vanadium toxicity in man are primarily respiratory. Aside from its acute inflammatory action on the lungs, it appears to exert its effect primarily on the various enzyme systems. Chronic exposure may depress cholesterol synthesis, oxidative phosphorylation in liver mitochondria will be uncoupled, and a decrease in urinary excretion of 5-hydroxyindoleacetic acid will occur with transient bilirubinuria and albuminuria. The condition is acute, with persistence of bronchospasm for 48–72 hr. This may indicate that vanadium oxide particles are present in the respiratory tract or that smooth muscle contraction is maintained by mechanisms other than direct chemical irritation.

Another symptom is the appearance of scattered eczematous skin lesions, some of which appear to be allergenic. These were found, for the most part, on exposed skin. The allergic phenomenon has been encountered in both workers and experimental animals; this included cutaneous eruptions, asthmatic bronchitis, and bronchial asthma. The increased cholinesterase activity is indicative of sensitization.

In view of the demonstrated absorption of vanadium and other observations, the possibility of chronic general poisoning must be considered. Among the persistent complications, the slight to moderate chronic changes in the mucous membranes of the upper respiratory tract (particularly the pharynx) should be noted, although no severe chronic bronchitis or changes in the lung have been reported. Several features— such as weakness, neurasthenia, slight anemia, increased thymol turbidity, increased sedimentation rate, and possibly allergy—may support the thesis of chronic systemic intoxication. This thesis has not been borne out by any conclusive evidence; if it does occur, it must be considered mild and of no great significance. It is of special importance that permanent damage to the target organs, including the lungs, has never been conclusively established and is doubtful.

Many of the findings reported in man have also been noted in animals, including changes in cardiovascular functions; reduction in urinary

5-hydroxyindoleacetic acid; in many cases, marked irritation of the
respiratory tract and conjunctiva; the presence of acute changes in the
lungs, which probably should be designated as "pneumonitis" principally of chemical–bacterial origin; mild, chronic inflammatory changes
in the respiratory tract; and the absence of specific chronic changes in
the lungs. There was, however, no clinical feature in man to correspond
to the emphysema, postpneumonic atelectasis, bronchopneumonic–
atelectatic changes, or enteritis seen in animals.

Several workers have been found to undergo an increase in sensitivity
after an initial heavy dose, whereas others seem to develop some adaptation to later exposures. Perhaps the term "hardening" or "inurement"
may apply to some men who appear more resistant than others to
chronic exposure to low concentrations of vanadium dusts.

It has been shown that vanadium has many metabolic effects. These
appear to include lowering of cholesterol synthesis (particularly in
younger animals and man), lowering of cystine and cysteine synthesis,
reduction in coenzyme A and coenzyme Q production, and activation
of monoamine oxidase. It also appears that vanadium may prevent
dental caries and affect iron metabolism.

Vanadium salts are poorly absorbed in the intestine. Even when they
are ingested in a soluble form, only about 1% of a given quantity is absorbed. Vanadium that is absorbed is excreted rapidly, with about 60%
of a given absorbed quantity being excreted in the urine in 24 hr. Only
small amounts of vanadium are stored, and vanadium that enters bone
is mobilized more slowly than that from other tissues. Because of the
low absorption of vanadium and its relatively rapid excretion, it is less
toxic than other metals that do not have these characteristics.

Although no evidence of nutritional essentiality for man was found,
vanadium is known to be essential for laboratory animals and probably
will be shown to be essential for man also.

The voluminous literature reviewed failed to reveal any indication of
mutagenicity from exposure to vanadium or its compounds.

6

General Summary
and Conclusions

GENERAL SUMMARY

Vanadium occurs ubiquitously in the earth's crust and is present at a concentration of about 150 ppm. However, few ores contain highly concentrated vanadium compounds, most of them ranging from 0.1% to 1% vanadium pentoxide, although some may contain as much as 3%. For this reason, the extraction of vanadium is associated with some other valuable product, such as uranium from carnotite ores.

In general, vanadium in an ore is converted to a soluble compound by heating (roasting) with sodium chloride to about 850 C. The water-soluble vanadium compounds thus formed are leached with hot water, and the vanadium is precipitated as sodium metavanadate by the addition of acid to a pH of 1–3. The residues from such roasting and leaching may be heaped on the ground or used as landfill, in which case they are subject to rain and groundwater drainage. Although this has apparently not been studied, these tailings dumps might be a source of water pollution for many miles around.

Construction steels contain about 0.1% vanadium, and tool and die steels may contain 0.5–3% vanadium. These are not believed to constitute any significant problem to the environment, because the vanadium is tightly bound in the steel. However, the recovery of such steels, especially by the basic-oxygen-furnace process, may result in the production of large amounts of fume, which would be expected to contain some of the vanadium. It is estimated that 7 tons of vanadium are emitted to the atmosphere from such sources each year. This is such a small amount that it is not believed to constitute a problem, except in immediate proximity to such operations. The magnitude of the potential local problem has not been fully studied.

In the production of ferrovanadium for alloy additions in steelmaking, the vanadium emission to the atmosphere has been estimated at 144 tons in 1968. Again, there is a possibility of a circumscribed vanadium problem, whose magnitude has not been fully evaluated in the vicinity of such plants.

Various vanadium compounds are used as catalysts in polymerization processes. When such polymers are used for food packaging or pharmaceutical and medical applications, a potential leaching hazard exists. This hazard has probably been controlled by the simulated food-leaching experiments required by the Food and Drug Administration before it grants approval for the use of a polymer in connection with foods or drugs.

The largest contributions to atmospheric pollution by vanadium are from the combustion of fossil fuels, especially coal and residual oils. The contribution of vanadium to the atmosphere in 1968 from coal combustion was estimated to range from 1,750 to 3,760 tons, depending on the basis of the estimates. The contribution of vanadium to the atmosphere from residual-fuel combustion was estimated at 12,400–19,000 tons in 1969 and 14,000–22,000 tons in 1970. Additional unknown quantities of vanadium are emitted to the atmosphere from the uncontrolled burning of nearly 300 coal-waste piles in the United States. Probably insignificant quantities of vanadium may be emitted to the atmosphere from the burning of wood (including forest fires), vegetable matter (agricultural wastes), and solid wastes. Vanadium has not been detected (sensitivity, better than 0.05 ppm) in distillate fuels, such as jet fuels, kerosene, gasoline, home heating oils, and automotive diesel fuels; hence, the combustion of these fuels is not believed to make any significant contribution to atmospheric vanadium.

The particle size distribution of the vanadium compounds emitted from the combustion of coal and oil is not known; thus, it is not known whether these vanadium compounds fall out near the source of combustion or remain suspended for considerable periods and distances. Furthermore, the ultimate fate of this vanadium is not known—that is, its half-life in the atmosphere, whether it is washed out by rain, and whether, in the low concentrations found in the ambient air (about 0.5–1.0 $\mu g/m^3$), it can catalyze oxidations, such as that of sulfur dioxide to sulfur trioxide.

It is anticipated that the increased demand for low-sulfur fuels, especially residual oils, will result in a significant reduction in vanadium emissions from the burning of residual oils. This is because, as residual oils are desulfurized, their vanadium content is reduced. Thus, in one reported case, a reduction in the amount of sulfur in fuel from 2.6% to 0.5% lowered the vanadium content of the fuel from 218 ppm to 59 ppm.

In air, vanadium concentrations range from as low as about 0.1 ng/m^3 to about 1 μg/m^3 (occasionally higher). In remote areas of the world, vanadium in the air seems to arise mainly from marine spray or continental dust. However, in both rural and urban areas in the eastern United States, the vanadium concentration in the air is much higher than continental dust would contribute and is probably related to air contamination by the ash from burned coal and residual-oil fuels.

Vanadium is present in the soil in concentrations ranging from less than 20 ppm to several hundred parts per million. It is present in seawater in concentrations of 0.002–0.029 ppm. Its concentration in plants varies widely, from undetectable to as high as 4 ppm in alfalfa. Its concentration in animal tissues ranges from less than 0.25 ppm to slightly over 1 ppm. Vanadium in drinking water ranges from undetectable to 220 ppb. In studies done over a period of years in several large cities, the atmospheric vanadium concentrations have shown no consistent upward or downward trend. However, in many cases, the vanadium data on plants and animal tissues are rather scanty and perhaps confused by analytic errors. It is apparent that far more analytic data, using the best analytic procedures, need to be obtained. Interestingly, some tunicates seem to be able to accumulate vanadium in concentrations up to 1,000 ppm. They seem to incorporate this vanadium into special cells, vanadocytes.

Vanadium has recently been shown to be an essential element for chicks and rats. Low concentrations of vanadium in the environment (in the parts-per-billion range) are probably beneficial, and even essential.

In high doses, vanadium is known to be toxic to animals. Thus, cats tolerate inhalation exposures to vanadium (as vanadium pentoxide) of 43 mg/m^3, but a concentration of 500 mg/m^3 caused death after 23 min of exposure. Experiments in rabbits showed that exposures to vanadium pentoxide at 20–40 mg/m^3 for 1 hr/day for several months produced lung and upper respiratory tract inflammation. Rats exposed at about the same concentration (10–30 mg/m^3) underwent similar changes. The toxicity of different vanadium compounds is highly variable; ferrovanadium (containing 30–85% vanadium) was tolerated up to a concentration of 2,000 mg/m^3. Dogs, rats, guinea pigs, and rabbits tolerated exposures to vanadium pentoxide at 0.5 mg/m³ for 6 hr/day for 6 months without change. Vanadium trioxide at 40–70 mg/m³ for 2 hr/day for 9–12 months produced a variety of changes in rabbits and rats. Vanadium chloride appeared more toxic than the trioxide.

Sodium metavanadate at 10–25 μg/g of diet caused adverse effects in chicks and rats, and vanadium (as vanadium pentoxide) at 10 μg/g in the diet caused a reduction in cystine content of the hair of rats and dogs.

When administered parenterally, different vanadium salts had different acute lethal doses, and species variations in toxicity were found among rabbits, guinea pigs, rats, and mice.

Vanadium appears to exert its effect—other than acute inflammation in the lungs when inhaled in high concentrations—through various enzyme systems. Thus, cholesterol synthesis is depressed, oxidative phosphorylation in liver mitochondria is uncoupled, and urinary excretion of 5-hydroxyindoleacetic acid is decreased, with transient bilirubinemia and albuminuria.

In man, the principal toxic effects of vanadium have been observed in industrial workers exposed on the job to relatively high concentrations of various vanadium compounds. In general, these exposures have been to several milligrams per cubic meter or higher. In these circumstances, direct irritation of the tracheobronchial tree results, with cough, sputum, tightness in the chest, wheezing, and eye, nose, and throat irritation. Some workers thus exposed also exhibited weakness, neurasthenia, and slight anemia, which suggest chronic toxic effects from absorption. Several workers have demonstrated an apparent increased sensitivity to vanadium compounds after a rather heavy initial exposure, whereas other workers seem to develop some adaptation to exposure.

In deliberate experimental exposure of volunteers to vanadium pentoxide at 0.2 mg/m³, loose productive cough was produced after an 8-hr exposure. Vanadium concentration in the urine peaked 3 days after exposure at 0.012 mg/100 ml, but was not detectable a week after exposure. Some fecal excretion of vanadium occurred, but had disappeared after 2 weeks.

In the United States, threshold limit values of 0.5 mg/m³ for vanadium pentoxide dust or vanadates and 0.05 mg/m³ for vanadium pentoxide fume have been proposed. In Russia, the proposed limit for dust is 0.1 mg/m³.

It appears that vanadium in the concentrations found in air, food, and water constitutes no hazard to the population as a whole. The vanadium concentration in industrial workplaces should be controlled so that it is at or below threshold limit values. The environment of vanadium processing plants should be monitored, and those who live there for a long period should have their fingernails or hair monitored for cystine content. If this is found to decrease or if the environment is found to be excessively contaminated, they should undergo further medical evaluation.

CONCLUSIONS

With the anticipated vanadium reductions in residual fuel oils, as desulfurization becomes more intense and more universal, controls for ambient atmospheres will probably not be necessary. For localized situations—e.g., ferrovanadium plants—improved fume and dust collection may be necessary if environmental vanadium concentrations are found to be significantly increased.

The Panel found no documentation of effects of vanadium or its compounds on materials in the concentrations normally encountered.

7

Recommendations

1. Some additional studies of the localized air-pollution problem from vanadium resulting from the recovery of scrap steel are warranted.

2. A similar potential problem exists in the vicinity of ferrovanadium production plants and warrants study to define its magnitude.

3. Studies of the particle size distribution of vanadium compounds emitted to the atmosphere from the combustion of coal, of oil, and of residual fuel oil that contains additives are needed. Furthermore, studies of the residence time of vanadium in the atmosphere are needed, as well as of its ultimate fate.

4. Detailed studies of the uptake and distribution of vanadium from the soil by plants and of the influence of vanadium content of the soil and soil type on this process are needed.

5. Studies of the uptake of vanadium by plants from air-pollution sources are needed.

6. More complete analyses of vanadium content of foods of plant and animal origin are needed.

7. Studies to determine whether vanadium-contaminated air and water significantly increase the vanadium content of animal tissues used as foods are needed.

8. Studies should be done to differentiate beneficial from harmful or neutral chemical forms of vanadium.

9. Studies of possible toxic effects of vanadium and its compounds at concentrations approximating those in the environment should be undertaken—i.e., between 0.01 and 1 $\mu g/m^3$ of air and between 0.1 and 1.0 $\mu g/g$ of diet. These studies should be concerned particularly with the effect of vanadium on the respiratory system among persons with different degrees of susceptibility and states of health. They should also include relations to iron nutrition and metabolism.

10. The effect of vanadium on the rate of metabolic processes should be re-evaluated using recently developed analytic techniques. These studies should include the effect of vanadium on cholesterol, phospholipid, and triglyceride synthesis.

11. Studies of the essentiality of vanadium in human nutrition are needed. These studies should include an examination of the utilization in nutrition of various chemical forms of vanadium.

12. Studies of the toxicity to man of vanadium in combination with other metal fumes or dusts are needed to determine whether synergistic or potentiating effects occur.

13. Because vanadium is a valuable metal, with industrial uses that require substantial importation to meet demand, the technologic feasibility of extracting vanadium from residual oils or recovering it from stack effluents should be explored. If an economical method of extracting or recovering 60% of the vanadium emitted from the combustion of residual oils were developed, it could fill nearly all the vanadium needs in the United States projected for 1975.

14. Studies are needed on the catalytic role of airborne vanadium particles in the oxidation of other air pollutants, such as sulfur dioxide, carbon monoxide, and nitrous oxide.

Appendix A
Desulfurization of
Residual Fuel Oils

A variety of processes have been developed for reducing the sulfur content of heavy (residual) fuel oils. All those thus far commercialized are catalytic hydrodesulfurization processes, and all involve passing a mixture of hydrogen and the product to be desulfurized over a catalyst at relatively high pressure and temperature. During this process, some (but not all) of the sulfur (which is present in crude oils in the form of a variety of organic sulfur compounds, and not as elemental sulfur) is converted to hydrogen sulfide. The hydrogen sulfide is later separated from the unused hydrogen and converted to elemental sulfur, usually by the Klaus process, in which a portion of the hydrogen sulfide is burned to form sulfur dioxide and then hydrogen sulfide and sulfur dioxide are reacted over a catalyst to yield elemental sulfur and water.

One of the chief differences among the processes is related to the nature of the stock fed to the desulfurization unit. In essence, processes have been designed to handle one of the following three feeds:

1. *Atmospheric residua* These consist of all the materials in a crude petroleum remaining after the crude petroleum has been processed in a pipe still to remove all the fractions that boil at up to about 340 C at atmospheric pressure. The residua contain most of the metals originally present as impurities in the crude oils from which they were produced.

79

2. *Heavy vacuum gas oils*   These are products obtained by distilling atmospheric residua at reduced pressure (in a vacuum pipe still) to avoid thermal cracking of the products. The vacuum gas oils generally have distillation ranges between 340 and 565 C equivalent atmospheric pressure. The preponderance of the metallic impurities present in the original crude oil is now concentrated in the bottoms that remain after the vacuum distillation step, and only traces of metals are in the vacuum gas oils.

3. *Deasphalted atmospheric residua*   In these products, the asphalt contained in the atmospheric residua is precipitated out of solution by the use of selective solvents (e.g., propane). The preponderance of the metallic impurities remains in the asphalt in this type of operation, although complete separation of the metals is not attained.

In desulfurizing any of these basic types of feedstocks to lower the sulfur content of residual fuel oils, the vanadium content of the resulting fuel oil is coincidentally reduced (but not to zero) for the reasons outlined below:

1. When atmospheric residua are hydrodesulfurized, some of the metallic impurities are laid down on the catalyst and are discarded later with the catalyst. This results in a shortening of the catalyst life, and the shortening usually is proportional to the metal content of the residuum being processed. Accordingly, because catalyst life is an important economic factor in the overall desulfurization process, the desulfurization of atmospheric residua is usually attractive only with residua that contain low concentrations of metallic impurities.

2. When low-sulfur residual fuel oil is made by blending desulfurized heavy vacuum gas oil with undesulfurized vacuum pipe-still bottoms, not all the vacuum pipe-still bottoms can be blended back into the finished fuel oil. In actual practice, the percentage of the bottoms that can be accommodated in the finished fuel oil decreases as the maximal desired sulfur content of the finished fuel decreases. Thus, less vacuum still bottoms can be blended into a fuel oil in meeting a sulfur specification of 0.5% than in meeting a sulfur specification of 1.0%. Because the vacuum still bottoms will contain nearly all the vanadium in the original crude oil, this results in a lowering of the vanadium content of the finished fuel oil. The vacuum still bottoms that cannot be blended into low-sulfur fuel oils may be disposed of in a number of ways, such as production of asphalt and production of fuel oils for markets where sulfur content is not critical.

3. A portion of the vanadium remaining in a deasphalted atmospheric

residuum will be deposited on the desulfurization catalyst, as explained earlier for undeasphalted atmospheric residuum. In this case, then, the vanadium content of the finished fuel oil will be reduced by the sum of the amount removed in deasphalting and the amount laid down on the catalyst.

At present, no commercial units designed specifically for lowering the sulfur content of residual fuel oils are being operated within the petroleum industry in the United States. (Many units are operating for hydro-desulfurizing light distillate products, such as naphthas, heating oils, and automotive diesel fuels.) Units for reducing the sulfur content of residual fuel oils are being operated, however, in the Caribbean area (e.g., in Venezuela and Aruba), and the products are being exported primarily to the East Coast of the United States. These units are all of the type that blend desulfurized vacuum gas oil with undesulfurized vacuum still bottoms.

The only units currently desulfurizing atmospheric residua are in Japan; one small unit is being planned for installation in Sweden.

No commercial units are operating on deasphalted atmospheric residuum. This appears to be related to the economics of this process, which are not as attractive as those of the other two processes described.

Appendix B
Detection and Measurement
of Vanadium in Biologic
and Pollution Materials

The various analytic techniques for measuring vanadium in biologic or pollution materials are reviewed in this appendix. This survey of methods is not intended to be comprehensive, but is meant to represent the recent trends in analytic methodology. Athanassiadis[9] has collected relevant information on analytic methods for vanadium, which was helpful as background for this discussion. Some general information is also presented on quantitative procedures; this information is related to the accuracy of analysis, which, unlike precision (repeatability), is not easily established.

A concern with accuracy is vital to the researcher in biology and pollution. The importance of the subject is too often brought into sharp focus when one attempts to compare analytic results among techniques and among laboratories. An example of the confusion that results from inadequate concern with accuracy appears in Table 4-14 (p. 44) and has to do with the vanadium content of food plants. The wide variations in vanadium content clearly indicate the measurement difficulties in some analytic procedures. Obviously, the question of proof of accuracy is critical, if analyses are to be interchangeable among laboratories. Interchangeability of results is essential in work concerned with establishing threshold limits of pollutants, defining background concentrations, and defin-

82

ing degrees of toxicity. Although it is seldom feasible to offer objective proof of accuracy for every method of analysis, there is little question but that careful application of well-established procedures can make it possible to improve efficiency in cooperative work on the biologic effects of atmospheric pollutants.

MEASUREMENT

Few analytic procedures applied to trace metals in biologic and pollution materials have been subjected to thorough, or even adequate, error analysis to permit rigorous definition of limits of accuracy. There are several reasons for this. It takes considerable effort to define sources of error in analytic procedures and to place limits on them. In addition, much of this work has been directed toward studies of concentration trends whereby repeatability of the analyses is essential, but proof-of-analysis accuracy is not deemed worth the effort.

If rigorously defined, accuracy requires not only that all sources of significant systematic error be identified and measured, but also that the analytic system be in statistical control, as defined by Natrella.[133] The General Test Methods of the American Society for Testing and Materials (ASTM)[7] contains definitions of the terms "precision" and "accuracy" and methods for their estimation in physical measurements. Every analytic procedure cited in the present report, for practical reasons, compromises the ideal in some way. However, as will be discussed, reasonable quantitative validation after the use of well-established procedures can result in useful interchange of analytic results among techniques and also among laboratories.

The various measuring procedures discussed here are generally applicable to all analytic techniques, but especially to those requiring calibrations with reference standards. The more common methods of measurement available to the analyst are listed below in approximate order of preference (within one or two rank positions)—i.e., method 1 is least subject to inaccuracies, and method 7 is most subject to inaccuracies.

1. Use of standard reference materials certified by recognized standardizing agency or by industrial suppliers of specific materials

2. Cooperative analyses involving several laboratories and several techniques (round robin)

3. Absolute analyses based on theoretical mathematical relations

4. Method of standard additions using solutions

5. Synthesized standards using solutions
6. Same as method 4, but using blended powders
7. Same as method 5, but using blended powders

In addition to these seven, there are in use a family of radiometric techniques,[124] of which the isotope-dilution method is one example. Although these methods are not truly quantitation procedures, they are important in this context, because they provide highly useful means for minimizing some analytic inaccuracies.

The unavailability of certified standard samples for trace metals in biologic and pollution materials precludes the use of method 1 in most cases. The National Bureau of Standards either has issued or is planning preparation of some biologic materials certified for trace-metal content, including freeze-dried bovine liver (SRM* 1577), tomato leaves (SRM 1573), orchard leaves (SRM 1571), tuna (SRM 1591), citrus leaves, alfalfa, pine needles, and aspen chips. Vanadium is not among the elements certified. Furthermore, the availability and long-term preservation of standards applicable to the broad range of matrices required for biologic and pollution samples will be very limited in the near future.

Method 2 requires a great deal of time, effort, and expense. It is most often applied to materials with great economic or social importance. This approach provides the unique opportunity to establish error limits under more realistic conditions than is feasible in a single laboratory. Groups currently working in cooperative sampling and analysis of atmospheric pollutants include the Intersociety Committee on Methods of Air Sampling and Analysis,[189] the ASTM Project Threshold,[8] and the Environmental Protection Agency.[208] However, this cooperative work is not necessarily conducted primarily to establish accuracy. The round-robin approach is more often conducted to establish uniform operating practices in several laboratories and to minimize the bias between laboratories using specified analytic procedures. In most cooperative work, the accuracy of the specified method is presumed to be established before distribution of samples to the cooperating laboratories. Nevertheless, a well-conducted round robin can reveal sources of analytic bias that have a bearing on accuracy.

Method 3 is exemplified by analyses based on the proven applicability of such theoretical relations as Beer's law in colorimetry or atomic absorption, the Nernst equation in electrochemical procedures, and the Ilkovic equation in polarography. Analytic procedures that have been shown to follow such relations are generally more amenable to good

*Standard reference material.

measurement than completely empiric methods, provided that interferences are carefully defined.

The method of standard additions, method 4, is one of the more
powerful techniques for minimizing systematic errors in analysis. The
automation of this method has been described by Leiritie and Mattsson.[101] Shatkay[166,167] has presented mathematical analyses of the
method of standard additions and of a similar technique, the method
of successive dilutions, including a discussion of the assumptions and
limitations of these methods that are often overlooked in their application.

Methods 4–7 involve synthesis of standards by blending either solutions or powders. Standards made from solutions are preferred over
mixtures of solids. The achievable accuracy of this procedure depends
on the close simulation of the standards to the samples. The more
accurately the composition of the sample is known, the better the
simulated composition can be. The synthesis of solid standards is
widely used, especially in emission spectroscopy and spark-source
mass spectroscopy. This method is subject to uncertainties that are
exceedingly difficult to resolve. A major problem with solid materials
not previously treated by dissolution is that the physical forms of the
additive standard materials should be identical with the form of the
analyte* in the unknown sample. To illustrate the subtle sources of
error possible with this method, Nohe and Mitteldorf[135] cite an example of relative errors of up to 75% caused by analyzing impurities in
unsintered beryllium oxide, compared with a matrix of sintered beryllium oxide. Another example of an effect of this type is cited by Morrison,[124] who shows relative differences of up to 100% in a matrix of
gamma-alumina (Al_2O_3) versus alpha-alumina. Apparently, therefore,
much greater discrepancies can occur because of chemical differences
between standards and samples. Some of the matrix effects with this
method can be reduced by diluting the sample in a uniform matrix.
In spite of the highly utilitarian nature of these dilution methods, it
is questionable whether they can be classified as quantitative without
considerable supporting evidence as to their accuracy and their applicability to variations in sample matrix.

Any cursory survey of published analytic techniques will reveal that
most offer only minimal evidence for inferring accuracy. Because this
problem will undoubtedly persist, it is especially important for analysts
and researchers in the biologic effects of atmospheric pollutants to be
aware of the problems of proving accuracy and to avoid some of the

---

*The specific chemical element sought in analysis.

pitfalls in regard to measurement. This will be especially true in the interim before standardized procedures can be validated.

Reagent Purity and Blanks

The precise assessment of and correction for analytes in reagents and solid additive materials are critical in trace-metal analysis. Inadequate control of blanks might be the most common cause of systematic errors at the nanogram level. Yoe and Koch,[210] Zief,[215] and Wahler[203] have discussed this topic, including reagent storage, purification, volatilization, distillation, and contamination from crushing and blending of powders. Robertson[152] has surveyed trace-metal concentrations in glass and plastic containment materials, organic and inorganic reagents, wiping tissues, and other materials. None of these writers reported the detection of vanadium in any of the commonly used reagents or in plastic containment materials by the most sensitive detection techniques. Thus, at present levels of detectability, vanadium appears to be one of the least troublesome elements with respect to contamination. However, the situation is made less favorable by the fact that vanadium is also among the least concentrated elements in biologic and pollution materials. Nevertheless, avoidance of inadvertent contamination of samples from unsuspected sources demands careful control in the interests of accuracy at the lowest concentrations.

Sample Preparation

Considerations of sample preparation are important with respect to economy and accuracy of analyses. The ideal approach is to analyze the sample directly, with no pretreatment at all.[50,100,206,216] However, direct analysis is not always possible and might even be undesirable because of the difficulty in compensating for matrix effects through synthesis of standards. Nearly all analytic procedures for biologic specimens and most analyses of particulate matter collected on paper filters involve some form of sample preparation. The preparation usually involves mineralization through ashing of the specimen. This can be accomplished in a muffle furnace[35,99,123,211] at temperatures between 400 and 650 C; by digesting in hot acid mixtures, such as nitric and perchloric acids;[42,54,211] or by ashing in an electrically excited oxygen atmosphere with a so-called low-temperature asher (L T A).[59,68,85,97,194] The primary concerns in the ashing operations are the loss of metals by volatilization, metal contamination, and convenience of the procedure. Vanadium is lost to some extent in the redistillation of heavy gas oils,[56,64] and this suggests possible volatilization losses in ashing procedures. Vanadium contamination when ashing in a porcelain crucible has been reported.[81]

In recent years, the LTA method has gained pre-eminence over other ashing procedures for organic materials of all types.[68,74,85,97,194,206] It is superior from the standpoints of minimal contamination and ease of operation. Some biologic materials that are incompletely ashed in a muffle at 500 C for 48 hr or in nitric and perchloric acids[59] are completely ashed in the LTA at 200 C for 24 hr.[85] None of the references specifically mentions the possible volatilization of vanadium using the LTA. However, from data reported for elements whose volatilities are comparable with those of vanadium and vanadium compounds, it can be inferred that vanadium is not significantly lost under normal conditions using the LTA.

The ashing temperatures for this method are between 150 and 250 C, as measured by infrared pyrometry. However, some workers (for example, M. Darr, National Bureau of Standards, and W. A. Gordon, National Aeronautics and Space Administration) have observed glowing particles when ashing carbonaceous materials with the LTA. This is apparently caused by exothermic oxidation reactions, which might result in unsuspected volatilization losses. In trace-metal work, contamination of the samples can occur by backstreaming of the vacuum-pump oils. This has been observed in the case of chromium at subnanogram concentrations (M. L. Taylor, Aerospace Research Laboratories, Wright-Patterson Air Force Base, private communication). In spite of these possible sources of error, the LTA method will probably replace other methods of organic decomposition in the future for all but a few of the most volatile elements.

Metal-containing constituents are extracted from ashed materials with combinations of nitric, perchloric, sulfuric, and hydrochloric acids.[42,50,59] Acid digestion of airborne particles on filter papers or of ashes from filter papers invariably leaves an insoluble residue. This residue contains siliceous and other mineral compounds, in addition to free carbon and graphite. The possibility of analytic bias caused by insoluble constituents has not been well-defined. Insoluble vanadium borides, nitrides, and silicides might be present in particulate samples or might be formed on ashing of the samples.[50] The ashes can be pelletized with graphite to provide an electrically conducting sample for later analysis by either emission spectroscopy[15] or spark-source mass spectroscopy.[35,59,211] Freeze-dried biologic materials were pelleted with graphite in much the same way.[15] However, comparison of results reported on freeze-dried material with results based on ash weight, wet weight, or other sample forms introduces complications.[41] To avoid

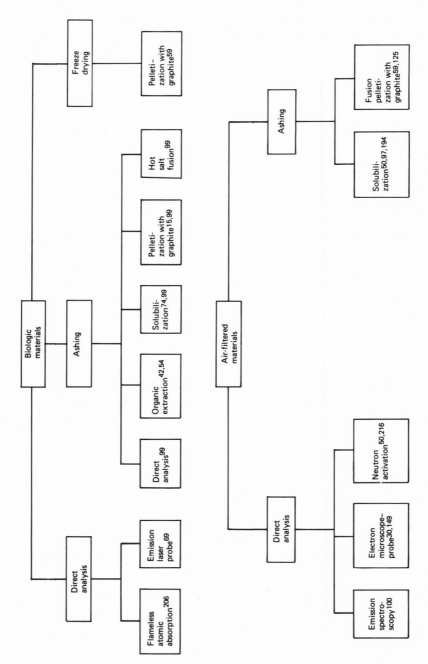

FIGURE B-1 Summary of sample preparation methods for biologic and air-filtered materials.

88

some of the uncertainties in sample preparation, the procedure whereby the ashes are solubilized by fusion appears to be advantageous.[125] Figure B-1 summarizes a variety of methods used in preparing biologic and pollution samples.

DETECTION

The methods for detecting vanadium in biologic and pollution materials have been based on considerations of detection limits or multielement capability, or both. The methods most prominent in the recent literature include neutron activation, emission spectroscopy, spark-source mass spectrometry, and atomic absorption. This section gives examples of application of these techniques to the detection of vanadium. Some special problems in applying these techniques to vanadium are also discussed. A unique and comprehensive study of airborne particles from six cities in the United States using a combination of many of the techniques described herein has been reported.[30] Other methods for vanadium, including chemical methods, are described in less detail. The special problem of sampling, which is both critical and complex, is left to other sources.[7]

Limits of Detection

In spite of the relatively high threshold limit values for vanadium compounds in industrial atmospheres, the problem of detecting ever smaller concentrations of vanadium remains important. The detection of small traces of vanadium is important in pollution work in establishing natural background concentrations and in using vanadium as a chemical tracer to indicate emission from oil-burning sources.[216] A typical filtered sample of urban air contains about 120 mg of particulate material per cubic meter, including as little as 50 μg of vanadium. Usually, only a small fraction of the total sample is available for vanadium analysis. In biologic work, the study and monitoring of vanadium toxicity requires the detection of small traces of vanadium in blood, urine, and tissues, to indicate possible exposure to relatively higher concentrations of vanadium compounds in the atmosphere. The characterization of blood serum is fundamental in the study of biologic systems; but there is currently no direct method for detecting vanadium in human serum. For practical work, therefore, the method of detection must have an absolute detection limit of about 1 ng/g and a relative detection limit of about 1 μg/g.

If reported detection limits are to have meaning, they must be carefully defined. A reasonably unambiguous definition may be made in terms of the experimental standard deviation of analytic results obtained at the blank concentration or at a very low concentration. The numerical limit is expressed at 1, 2, or 3 times the standard deviation. The integer selected is arbitrary, and for at least 11 replications it represents the approximate 70%, 90%, and 98% confidence limits, respectively. In atomic-absorption methods, the detection limit is almost universally defined as the amount of substance yielding 1% absorption. As a general rule, multiplying detection limits by 10 yields an estimate of the practical limit of detection.

Neutron Activation

Vanadium has been detected, with other elements, by direct irradiation of particulate aerosols collected on filter media.[50,216] Chemical separations were not necessary in these applications, because the concentrations of sodium and other interfering elements were low and because a lithium-drifted germanium [Ge(Li)] detector was used in the analysis of the radioactive species. This detector, a rather recent innovation, allows measurements of energy spectra with about 20 times better resolution than that of the more conventional thallium-activated sodium iodide [NaI(Tl)] detector, thus reducing the necessity of chemical separations of interferences.

In neutron-activation analysis of vanadium, the radioactive species produced is vanadium-52, with a half-life of 3.77 min and a gamma energy of 1.434 MeV.[108] The vanadium-52 can also be produced from chromium and manganese present in the sample by fast-neutron activation. These elements, therefore, constitute interference if they are present at high concentrations.

Measurement with this procedure requires comparison with a synthesized standard. Ideally, the standard should contain the same major elements as the sample, at about the expected concentrations and in the same containment geometry. However, because of total penetration of neutrons through the sample, the effects of chemical combination and physical form are negligible with this method. Detection limits by this procedure depend on sample composition. Limits of detection for paint, water, fish, and plastics are about 3.5, 0.015, 0.56, and 0.013 μg/g, respectively.[3] For air-filtered samples, the detection limit was about 1 ng,[50,216] corresponding to a concentration limit of about 2 ng/m^3 in typical urban air. The detection limit of vanadium in most biologic materials is complicated by the relatively high concentrations of sodium, even when using the Ge(Li) detector.[148] The sodium interference can be removed by absorption on hydrated antimony pentoxide.[148] However, this procedure is not applicable to the determination of vanadium in typical biologic materials because of the almost total decay of the short-lived vanadium-52 during the several hours required to process samples after irradiation. Removal of the sodium before irradiation might allow the detection of vanadium if blanks are properly controlled.

An alternative to matrix removal is removal of the vanadium from the biologic ash by solvent extraction. This approach was used to detect vanadium in foods at concentrations down to 2 μg/g.[99] The sample processing was completed in about 20 min, during which time the vanadium activity decayed to about 2.5% of the maximum.

Vanadium was determined in natural waters by neutron activation, by collecting vanadium on an ion-exchange resin and later irradiating the nitric acid effluent.[107] Concentrations of vanadium typically ranged between 1 and 10 μg/g in some rivers of the southwestern United States.

Emission Spectroscopy

Emission spectrochemical methods are differentiated by method of sample preparation, method of quantitation, and method of exciting the atomic spectra. As with instrumental methods of analysis in general, a wide variety of experimental procedures have been developed that represent tradeoffs between simplicity and economy on the one hand, and precision and accuracy on the other. One of the simplest methods of analyzing air-filtered samples by emission spectroscopy requires no pretreatment of the sample at all.[100] Filter papers 1 in. square are rolled and sparked directly, using a novel technique to push the paper into the analytic gap. Although the repeatability of the method is adequate for monitoring concentration trends of some elements, its accuracy is unknown, because calibration standards have consisted of dried solution of the various elements on filter paper. Quantitation of the method using air-filtered specimens analyzed by other means is a possible variation.

Another procedure for detecting vanadium and nine other elements in suspended particulate matter has been reported.[125] In this method, the chemical forms of the particles were destroyed by ashing and then fusing the ash with lithium tetraborate. The ground fusion material was then pelletized with graphite and subjected to a spark discharge. This technique tends to minimize inaccuracies caused by differences in chemical form between unknown samples and standards, assuming that the composition of the major elements in the standards is approximately the same as that in the samples. The mean vanadium concentration in metropolitan New York air was found to be 0.17 μg/m^3 using this method.

Possibly the most often used emission spectroscopic procedure for the detection of metals in airborne particulate materials and in biologic materials is a variation of the so-called universal method of analysis.[31] Many commercial laboratories use this general approach because of its economic attractiveness. There are many variations of this approach, which basically involves diluting the ashed sample in a relatively pure powdered material in a sample : diluent ratio of about 1 : 10. Typical diluent materials are graphite, lithium carbonate, and lithium fluoride. The dilution reduces all samples to a relatively common matrix and therefore reduces some systematic errors caused by variation in sample composition. However, there remains in these procedures a fundamental uncertainty concerning physical forms of the various chemical species, as discussed earlier. Because it is not possible to place limits on this source of error, these methods require quantitative validation by other means.

General emission spectroscopic methods for biologic materials include a method based on the dilution procedure described above,[196] a comprehensive analysis using freeze-dried materials,[15] and a procedure in which excitation is performed in a gastight arc chamber.[74] In addition, a specialized emission procedure using a laser to detect trace metals in small local areas on a cellular scale has been reported.[69,111] Although these methods are not specifically optimized to detect vanadium, they are all applicable to its detection and are therefore discussed briefly here. Tipton et al.[196] used graphite as the diluent and quantitated by addition of metal compounds to a synthetic matrix simulating biologic ashes. Bedrosian et al.[15] used graphite as the diluent and quantitated by additions of metal oxides to the freeze-dried matrix and to a synthetic biologic matrix consisting of p-nitrobenzeneazoresorcinol. No attempt was made in either of these methods to define possible inaccuracies caused by physicochemical differences between the standards and the samples. In the method of Hambidge,[74] the biologic ash was solubilized in dilute hydrochloric acid, and analyte elements were added either before or after ashing. Reagent blanks were kept to a minimum by using only 10 μg of acid reagent per sample. The sample solution was micropipetted onto a carbon electrode that contained 4 mg of silver chloride. The solutions added to the electrode were dried, and the residue contained in the porous electrode was arced in an atmosphere of argon. The detection limit is about 1 ng/g for vanadium, or about 0.5 μg/g in biologic ashes. Measurement was accomplished by adding known amounts of analyte elements to liquid samples (serum) and to solid samples (hair) before ashing. The method of standard additions was used for measurement.

The laser method of exciting atomic spectra provides the method of measuring metal constituents in biologic tissues *in vivo*. In addition, the determinations can be made on areas as small as 5 μm in diameter and 1–3 μm deep. Measurement is difficult, in view of the difficulty of defining sample volumes and matrix effects. Vanadium has not been reported in body fluids or tissues with this procedure, in spite of reported detection limits of about 10^{-15} g.[69]

Spark-Source Mass Spectrometry

The spark-source mass spectrometer has been used for multielement analysis of both airborne particles and biologic samples.[59,99,211] The sample preparation in all cases consisted of ashing the sample and pelletizing the ash with graphite to achieve the necessary electric conductivity. Trace elements in human hair were the subject of the work of Yurachek *et al.*,[211] but vanadium was not among the elements detected. Evans and Morrison[59] reported some general problems using this technique for biologic ashes. First, the ashing must be complete because of the numerous interferences that are otherwise produced by organic species. Second, vanadium was in a class of elements of which inorganic species commonly found in biologic materials interfered with all vanadium isotopes except one. Therefore, the possibility of isotopic interferences could not be eliminated by measuring isotopic ratios, and the vanadium concentrations determined represent upper limits only. Vanadium concentrations of two lung specimens were determined to be 0.33 and 12 μg/g and were more than 10 times higher than those determined by emission spectroscopy.[59]

A more recent innovation is the use of electric detection with the spark-source mass spectrometer. This new method promises to simplify the technique and to improve the detection repeatability.[27]

By the mass-spectrometric technique, the vanadium content of New York City air was determined[35] to be 1.9 μg/m^3. The particulate sample was collected on nitrocellulose filters that were ashed at 450 C and pelletized with graphite. No account was taken of relative sensitivity factors, nor of metal losses in the ashing step.

Atomic Absorption

The atomic-absorption technique is basically a single-element method but is nevertheless often advantageous in the measurement of trace metals because of its wide availability and its relative simplicity. Vanadium forms thermally stable oxides that are only partially dissociated in the flame. Therefore, the hottest flame in common use, the nitrous oxide–acetylene flame, is used to achieve the lowest detection limits. Although the lower detection limit is achieved in the emission mode,[42] the absorption mode has been more often applied to the materials discussed here.

There are two basic embodiments of the atomic-absorption mode— the flame–aspiration procedure and the flameless furnace–vaporization technique. The flame–aspiration method is preferred, where applicable, because it is easier to quantitate, simpler in operation, and more readily reproducible. The furnace technique is advantageous for analyzing solid samples that cannot be easily solubilized and in cases where the flame does not provide sufficient sensitivity.

The atomic-absorption method for detecting vanadium in ores[70] was adapted by the Intersociety Committee on Methods of Air Sampling and Analysis and specified as a Tentative Standard Method for vanadium in air-filtered samples.

Flame atomic absorption was applied to the detection of metals in blood after solvent extraction with methylisobutylketone.[42] However, vanadium was not included among the metals detected. Characteristic problems caused by aspiration of organic compounds typical of such extraction procedures are discussed by Delves *et al.*[54]

Microgram quantities of vanadium in lake water were detected by atomic absorption after extraction with 5,7-dichloro-8-hydroxyquinoline.[40] The atomic-absorption method was also described[123] for detecting metals in airborne particles. The sample, collected on glass-fiber filters, was ashed either in a muffle furnace at 550 C or in an LTA. The filter was digested with redistilled nitric acid, and the acid solution was aspirated into the flame. There were no apparent problems with nonquantitative extractions of vanadium. The limit of detection was 0.0018 μg/m^3 for a total air volume of 500 m^3 in 8.9 ml of solvent, or about 0.1 μg/ml.

Kneip *et al.*,[97] in another application of atomic absorption to airborne particles, reported vanadium concentrations of 0.115 μg/m^3 in a nonurban area of New York and 1.46 μg/m^3 in the Bronx. The detection limit for vanadium was reported as 0.094 μg/m^3 for a total air volume of 5,000 m^3. This is equivalent to a relative detection limit of about 10 μg/ml.

The recent development of the flameless atomic-absorption method as a routine laboratory tool was motivated primarily by analytic needs in biologic and pollution applications. In this method, the specimen is thermally vaporized in nonair atmosphere inside a graphite cell[136,206] or on a metal strip.[89] Vanadium at about 0.4 ng/m^3 can be detected under interference-free conditions. The relative sensitivity in micrograms per gram depends on the type of sample analyzed.

Vanadium was detected in mineral oils by flameless atomic absorption by adding 10-μl samples to the graphite tube.[136] The sensitivity was about 0.2 μg/g of oil. Concentrations of vanadium in mineral oils ranged from less than 0.2 μg/g in oil from Nigeria to 225 μg/g in oil from Venezuela.

The flameless method was also applied to the detection of various metals in biologic materials, including whole blood, without sample pretreatment.[206] However, no one has yet reported the detection of vanadium in biologic or pollution materials by any flameless method.

Colorimetry

Numerous colorimetric methods for detecting vanadium have been described. These are generally applicable to pollution and biologic materials with the use of suitable masking agents and extraction procedures to separate vanadium from interfering species. Some recently described colorimetric procedures, not necessarily applied to pollution or biologic materials, are listed in Table B-1. The 8-hydroxyquinoline procedure[190] was adapted to air-filtered materials and specified as a tentative method by the Intersociety Committee on Methods of Air Sampling and Analysis. The 8-hydroxyquinoline reagent was also applied after extraction of vanadium with α-benzoinoxime in chloroform.[119]

Electrometric Methods

Methods based on electrolytic phenomena are highly diverse in application and include at least 13 distinct techniques. These methods have found very limited use in the detection of vanadium in biologic or pollution materials.

The area of most active investigation today is that of anodestripping voltametry (ASV), which provides the advantages of preconcentration, reasonably good specificity, sensitivity, and simplicity. However, vanadium is one of the more difficult elements to detect by ASV because of the lack of a suitable reversible reaction. Therefore, the detection of vanadium in the materials of interest by this technique has not yet been reported. Vanadium has been detected by other electrometric techniques, including polarography,[92] potentiometry,[38] amperometry,[168,169] and coulometry.[98,150]

TABLE B-1 Colorimetric Reagents for Detecting Vanadium

Reagent	Reference	Remarks[a]
8-Hydroxyquinoline	190	—
Cyclo-tris-7-(1-azo-8-hydroxy)naphthalene-3,6-disulfonic acid (calcichrome)	90	—
N-o-Tolylbenzoylhydroxamic acid	110	—
m-Nitro-N-phenylbenzoylhydroxamic acid	67	Beer's law followed from 0.2 to 11 μg/g
Unsaturated N-arylhydroxamic acids (23 complexes studied)	26	Nine complexes had $\epsilon \simeq 1,500$
N-Phenylbenzohydroxamic acid	26	$\epsilon = 4,650$
N-Phenyl-3-styrylacrylohydroxamic acid	26	Beer's law followed from 0.7 to 8.4 μg/g; sensitivity, 0.0068 μg/cm^2
4-(2-Pyridylazo)resorcinol	170	—
Diaminobenzidine	39	—
5-Amino-4-hydroxy-3-(2-hydroxy-2,5-dinitrophenylazo)naphthalene-2,7-disulfonic acid (picraminazo-N)	212	$\epsilon = 12,400$
5-Amino-3-(3-chloro-2-hydroxy-5-nitrophenylazo)-4-hydroxynaphthalene-2,7-disulfonic acid (gallion)	212	$\epsilon \simeq 8,700$
Naphthalene-2,3-diol (2,3-dihydroxynaphthalene)	137	—

$^a\epsilon$, extinction coefficient.

Electron Optics

Electron optics includes the electron microscope, the electron microprobe, and x-ray diffraction. These tools, supplemented by light microscopy, allow characterization beyond elemental analysis and into morphologic, compound, and crystallographic identification, as described by Rhoads[149] and Blosser.[30] Although no vanadium compounds have yet been identified with these techniques, vanadium distributions within particulate-material samples in the Washington, D.C., area have been reported.[30]

X-Ray Fluorescence

When applicable, x-ray fluorescence is convenient, precise, and relatively easy to measure. It has been used to detect vanadium in fuel oils,[52,57,93] biologic materials,[5] and particulate material filtered from the air.[30] The limit of detection in aqueous solutions has been reported as 0.5 μg/ml[109] and 1.5 μg/ml.[23] The detection limit in sodium tetraborate fusion was about 3 μg/ml.[23]

Gas Chromatography

Hyperpressure gas-phase chromatographic separations of organic vanadium compounds and their later detection have been described.[94,126] This method is emerging from the developmental stage and has found practical application to trace metals in biologic materials.[192] However, no practical application has been reported for vanadium.

References

1. Abernethy, R. F., and F. H. Gibson. Rare Elements in Coal. Bureau of Mines Report IC-8163. Washington, D.C.: U.S. Department of the Interior, 1963. 73 pp.
2. Abernethy, R. F., M. J. Peterson, and F. H. Gibson. Spectrochemical Analyses of Coal Ash for Trace Elements. Bureau of Mines Report of Investigations 7281. Pittsburgh, Pa.: U.S. Department of the Interior, 1969. 30 pp.
3. Activation Analysis. A Summary. San Diego: Gulf Radiation Technology. (no date) 6 pp.
4. Aiyar, A. S., and A. Sreenivasan. Effect of vanadium administration on co-enzyme Q metabolism in rats. Proc. Soc. Exp. Biol. Med. 107:914–916, 1961.
5. Alexander, G. V. X-ray fluorescence analysis of biological tissues. Appl. Spectrosc. 18:1–4, 1964.
6. American Conference of Governmental Industrial Hygienists. Documentation of the Threshold Limit Values for Substances in Workroom Air. (3rd ed.) Cincinnati, Ohio: American Conference of Governmental Industrial Hygienists, 1971. 286 pp.
7. American Society for Testing and Materials. 1971 Annual Book of ASTM Standards. Part 30. Philadelphia: American Society for Testing and Materials, 1971. 1450 pp.
8. American Society for Testing and Materials. Project Threshold. The ASTM D-22 Program to Validate Standard Test Methods for Implementing Air Pollution Control. Progress Report. Philadelphia: American Society for Testing and Materials, 1971. 15 pp.
9. Athanassiadis, Y. C. Air Pollution Aspects of Vanadium and Its Compounds. (Prepared for the National Air Pollution Control Administration by Litton

Systems, Inc., Bethesda, Md.) Springfield, Va.: Clearinghouse for Federal Scientific and Technical Information, 1969. 93 pp.

10. Athanassiadis, Y. C. Preliminary Air Pollution Survey of Vanadium and Its Compounds. A Literature Review. National Air Pollution Control Administration Publication APTD 69-48. Springfield, Va.: Clearinghouse for Federal Scientific and Technical Information, 1969. 91 pp.

11. Azarnoff, D. L., F. E. Brock, and G. L. Curran. A specific site of vanadium inhibition of cholesterol biosynthesis. Biochim. Biophys. Acta 51:397–398, 1961.

12. Azarnoff, D. L., and G. L. Curran. Site of vanadium inhibition of cholesterol biosynthesis. J. Amer. Chem. Soc. 79:2968–2969, 1957.

13. Barannik, P. I., I. A. Mikhalyuk, I. N. Motuzkov, and G. S. Yatsula. Levels of trace elements and natural radioactivity of food products of some areas of the Kiev region. Vopr. Pitan. 29(1):79–81, 1970. (in Russian)

14. Beard, H. H., R. W. Baker, and V. C. Myers. Studies in the nutritional anemias of the rat. V. The action of iron and iron supplemented with other elements upon the daily reticulocyte, erythrocyte, and hemoglobin response. J. Biol. Chem. 94:123–134, 1931.

15. Bedrosian, A. J., R. K. Skogerboe, and G. H. Morrison. Direct emission spectrographic method for trace elements in biological materials. Anal. Chem. 40:854–860, 1968.

16. Belgian Patent 543292 (Goodrich–Gulf Chemicals). Perfectionnements Apportés à la Production de Polymères et de Copolymères d'Hydrocarbures Polyolefines Conjugués, 1956.

17. Berg, L. R. Effect of diet composition on vanadium toxicity for the chick. Poult. Sci. 45:1346–1352, 1966.

18. Berg, L. R. Evidence of vanadium toxicity resulting from the use of certain commercial phosphorus supplements in chick rations. Poult. Sci. 42:766–769, 1963.

19. Berg, L. R., G. E. Bearse, and L. H. Merrill. Vanadium toxicity in laying hens. Poult. Sci. 42:1407–1411, 1963.

20. Bergel, F., R. C. Bray, and K. R. Harrap. A model system for cysteine desulphydrase action: Pyridoxal phosphate–vanadium. Nature 181:1654–1655, 1958.

21. Bernheim, F., and M. L. C. Bernheim. Action of vanadium on tissue oxidations. Science 88:481–482, 1938.

22. Bernheim, F., and M. L. C. Bernheim. The action of vanadium on the oxidation of phospholipids by certain tissues. J. Biol. Chem. 127:353–360, 1939.

23. Bertin, E. P. Solution techniques in x-ray spectrometric analysis. Norelco Reporter 12(1):15–26, Jan.–Mar. 1965.

24. Bertrand, D. Recherches sur le vanadium chez les vegetaux. Bull. Soc. Chim. Biol. 23:391–397, 1941.

25. Bertrand, D. Survey of contemporary knowledge of biogeochemistry. 2. The biogeochemistry of vanadium. Bull. Amer. Museum Nat. Hist. 94:403–455, 1950.

26. Bhura, D. C., and S. G. Tandon. Unsaturated N-arylhydroxamic acids as colorimetric reagents for vanadium (V). Spectrophotometric determination with N-phenyl-3-styrylacrylhydroxamic acid. Anal. Chim. Acta 53:379–386, 1971.

27. Bingham, R. A., and R. M. Elliot. Accuracy of analysis by electrical detection in spark source mass spectrometry. Anal. Chem. 43:43–54, 1971.

28. Bituminous Coal Facts. Washington, D.C.: National Coal Association, 1970. 90 pp.

29. Blankenhorn, D. H., and H. P. Chin. Management of hypercholesterolemia. GP 29:134–142, 1964.

30. Blosser, E. R. Identification and Estimation of Ions, Molecules, and Compounds in Particulate Matter Collected from Ambient Air. PB-201 738. Final Report Prepared for the Environmental Protection Agency, Air Pollution Control Office. Contract CPA-70-159. Springfield, Va.: Department of Commerce, National Technical Information Service, 1971. 77 pp.

31. Boumans, P. W. J. M. Theory of Spectrochemical Excitation, pp. 203–208. New York: Plenum Press, 1966.

32. Bowden, A. T., P. Draper, and H. Rawling. The problem of fuel oil deposition in open-cycle gas turbines. Proc. (A) Inst. Mech. Engr. 167:291–313, 1953.

33. Bowen, H. J. M. Trace Elements in Biochemistry. New York: Academic Press, Inc., 1966. 241 pp.

34. Brar, S. S., D. M. Nelson, J. R. Kline, P. F. Gustafson, E. L. Kanabrocki, C. E. Moore, and D. M. Hattori. Instrumental analysis for trace elements in Chicago area surface air. J. Geophys. Res. 75:2939–2945, 1970.

35. Brown, R., and P. G. T. Vossen. Spark source mass spectrometric survey analysis of air pollution particulates. Anal. Chem. 42:1820–1822, 1970.

36. Browne, R. C. Vanadium poisoning from gas turbines. Brit. J. Ind. Med. 12:57–59, 1955.

37. Browne, R. C., and J. Steel. The control of the vanadium hazard in catalytic oil-gas plants. Ann. Occup. Hyg. 6:75–79, 1963.

38. Cassani, F. Determinazione potenziometrica del titanio e del vanadio. Chim. Ind. (Milan) 51:1248–1251, 1969.

39. Chan, K. M., and J. P. Riley. The determination of vanadium in sea and natural waters, biological materials and silicate sediments and rocks. Anal. Chim. Acta 34:337–345, 1966.

40. Chau, Y. K., and K. Lum-Shue-Chan. Complex extraction of vanadium for atomic absorption spectroscopy. Determination of microgram qualities of vanadium in lake waters. Anal. Chim. Acta 50:201–207, 1970.

41. Christian, G. D. Medicine, trace elements, and atomic absorption spectroscopy. Anal. Chem. 41:24A–40A, 1969.

42. Christian, G. D., and F. J. Feldman. Atomic Absorption Spectroscopy Applications in Agriculture, Biology, and Medicine. New York: Wiley–Interscience, 1970. 490 pp.

43. Ciereszko, L. S., E. M. Ciereszko, E. R. Harris, and C. A. Lane. Vanadium content of some tunicates. Comp. Biochem. Physiol. 8:137–140, 1963.

44. Colorado Vanadium—A Composite Study. Denver: Colorado State Metal Mining Fund, 1961. 155 pp.

45. Cuffe, S. T., and R. W. Gerstle. Emissions from Coal-Fired Power Plants: A Comprehensive Summary. Public Health Service Publication 999-AP-35. Cincinnati: Public Health Service, National Center for Air Pollution Control, 1967. 30 pp.

46. Curran, G. L. Effect of certain transition group elements on hepatic synthesis of cholesterol in the rat. J. Biol. Chem. 210:765–770, 1954.

47. Curran, G. L., D. L. Azarnoff, and R. E. Bolinger. Effect of cholesterol synthesis inhibition in normocholesteremic young men. J. Clin. Invest. 38:1251–1261, 1959.

48. Curran, G. L., and R. E. Burch. Biological and health effects of vanadium, pp. 96–104. In Proceedings. University of Missouri's 1st Annual Conference on Trace Substances in Environmental Health. Columbia: University of Missouri, 1967.

49. Curran, G. L., and R. L. Costello. Reduction of excess cholesterol in the rabbit aorta by inhibition of endogenous cholesterol synthesis. J. Exp. Med. 103:49–56, 1956.

50. Dams, R., J. A. Robbins, K. A. Rahn, and J. W. Winchester. Nondestructive neutron activation analysis of air pollution particulates. Anal. Chem. 42:861–867, 1970.

51. Daniel, E. P., and R. D. Lillie. Experimental vanadium poisoning in the white rat. Public Health Rep. 53:765–777, 1938.

52. Davis, E. N., and B. C. Hoeck. X-ray spectrographic method for the determination of vanadium and nickel in residual fuels and charging stocks. Anal. Chem. 27:1880–1884, 1955.

53. Davis, W. E., and Associates. National Inventory of Sources and Emissions. Arsenic, Beryllium, Manganese, Mercury and Vanadium. 1968. Vanadium. Section V. Report for Environmental Protection Agency. Leawood, Kansas: W. E. Davis and Associates, 1971. 53 pp.

54. Delves, H. T., G. Shepherd, and P. Vinter. Determination of eleven metals in small samples of blood by sequential solvent extraction and atomic-absorption spectrophotometry. Analyst 96:260–273, 1971.

55. Dimond, E. G., J. Caravaca, and A. Benchmol. Vanadium. Excretion, toxicity, lipid effect in man. Amer. J. Clin. Nutr. 12:49–53, 1963.

56. Dunlop, E. C. Decomposition and dissolution of samples: Organic, pp. 1051–1093. In I. M. Kolthoff and P. J. Elving, Eds. Treatise on Analytical Chemistry. Part I. Theory and Practice. Vol. 2. New York: Interscience Publishers, Inc., 1961.

57. Dwiggins, C. W., Jr., and H. N. Dunning. Quantitative determination of traces of vanadium, iron, and nickel in oils by x-ray spectrography. Anal. Chem. 32:1137–1141, 1960.

58. Eckardt, R. E. Petroleum fuel and airborne metals. Arch. Environ. Health 23:166–167, 1971. (letter to the editor)

59. Evans, C. A., Jr., and G. H. Morrison. Trace element survey analysis of biological materials by spark source mass spectrometry. Anal. Chem. 40:869–875, 1968.

60. Faulkner Hudson, T. G. Vanadium. Toxicology and Biological Significance. New York: Elsevier Publishing Co., 1964. 140 pp.

61. Fishburn, C. W., and C. Zenz. Metal fume fever. A report of a case. J. Occup. Med. 11:142–144, 1969.

62. Franke, K. W., and A. L. Moxon. The toxicity of orally ingested arsenic, selenium, tellurium, vanadium and molybdenum. J. Pharmacol. Exp. Therap. 61:89–102, 1937.

63. From Highveld slag to ferrovanadium at the GFE Nuremberg works. Highveld Vanadium News 2:4–5, 1971.

64. Gamble, L. W., and W. H. Jones. Determination of trace metals in petroleum. Wet ash-spectrographic method. Anal. Chem. 27:1456–1459, 1955.

65. Gaylord, N. G., and H. F. Mark. Linear and Stereoregular Addition Polymers: Polymerization with Controlled Propagation. New York: Interscience Publishers, Inc., 1959. 571 pp.

66. Geyer, C. F. Vanadium. A caries inhibiting trace element in the Syrian hamster. J. Dental Res. 35:590–595, 1953.
67. Ghosh, N. N., and G. Siddhanta. Extraction–photometric determination of vanadium(V) with N-(m-nitrobenzoyl)-N-phenylhydroxylamine. Fresenius' Z. Anal. Chem. 253:207–208, 1971.
68. Gleit, C. E., and W. D. Holland. Use of electrically excited oxygen for the low temperature decomposition of organic substances. Anal. Chem. 34:1454–1457, 1962.
69. Glick, D. Cytochemical analysis by laser microprobe–emission spectroscopy. Ann. N.Y. Acad. Sci. 157:265–274, 1969.
70. Goeke, R. Determination of vanadium in ore samples by atomic-absorption spectrophotometry. Talanta 15:871–873, 1968.
71. Gordon, G. E., W. H. Zoller, E. S. Gladney, and A. G. Jones. Neutron activation studies of trace elements on Boston-area atmospheric particles. Presented at American Chemical Society Meeting, Boston, Massachusetts, April 1972.
72. Hadjimarkos, D. M. Effects of trace elements on dental caries. Adv. Oral Biol. 3:253–292, 1968.
73. Hadjimarkos, D. M. Vanadium and dental caries. Nature 209:1137, 1966.
74. Hambidge, M. K. Use of static argon atmosphere in emission spectrochemical determination of chromium in biological materials. Anal. Chem. 43:103–107, 1971.
75. Harrison, P. R., K. A. Rahn, R. Dams, J. A. Robbins, J. W. Winchester, S. S. Brar, and D. M. Nelson. Area wide trace metal concentrations measured by multielement neutron activation analysis. One day study in Northwest Indiana. J. Air Pollut. Control Assoc. 21:563–570, 1971.
76. Hathcock, J. N., C. H. Hill, and G. Matrone. Vanadium toxicity and distribution in chicks and rats. J. Nutr. 82:106–110, 1964.
77. Hathcock, J. N., C. H. Hill, and S. B. Tove. Uncoupling of oxidative phosphorylation by vanadate. Can. J. Biochem. 44:983–988, 1966.
78. Heege, J. H. T. Poisoning of cattle by ingestion of fuel oil soot. Tijdschr. Diergeneesk. 89:1300–1304, 1964. (in Dutch)
79. Hein, J. W., and J. Wisotzky. The effect of a 10 PPM vanadium drinking solution on dental caries in male and female Syrian hamsters. J. Dental Res. 34:756, 1955.
80. Henze, M. Über das Vanadiumschromogen des Ascideinblutes. Hoppe–Seylers Z. Physiol. Chem. 213:125–135, 1932.
81. Heydorn, K., and H. R. Lukens. Pre-irradiation Separation for the Determination of Vanadium in Blood Serum by Reactor Neutron Activation Analysis. Report. RISO-138. Risoe: Danish Atomic Energy Commission, 1966. 20 pp.
82. Hickey, R. J., E. P. Schoff, and R. C. Clelland. Relationship between air pollution and certain chronic disease death rates. Multivariate statistical studies. Arch. Environ. Health 15:728–738, 1967.
83. Hoffman, G. L. Ph.D. Thesis. University of Hawaii, 1971.
84. Hoffman, G. L., R. A. Duce, and W. H. Zoller. Vanadium, copper, and aluminum in the lower atmosphere between California and Hawaii. Environ. Sci. Technol. 3:1207–1210, 1969.
85. Hollahan, J. R. Topics in chemical instrumentation. XXVII. Analytical applications of electrodelessly discharged gases. J. Chem. Educ. 43:A401–A404, 1966.
86. Hopkins, L. L., Jr., and H. E. Mohr. The biological essentiality of vanadium, pp. 195–213. In W. Mertz and W. E. Cornatzer, Eds. Newer Trace Elements in Nutrition. New York: Marcel Dekker, 1971.

87. Hopkins, L. L., Jr., and B. E. Tilton. Metabolism of trace amounts of vanadium[48] in rat organs and liver subcellular particles. Amer. J. Physiol. 211: 169–172, 1966.

88. Hulcher, F. H. Spectrophotometric determination of vanadium in biological material. Anal. Chem. 32:1183–1185, 1960.

89. Hwang, J. Y., P. A. Ullicci, and S. B. Smith, Jr. A simple flameless atomizer. Amer. Lab. 3:41–43, 1971.

90. Ishii, H., and H. Einaga. Use of calcichrome as a spectrophotometric reagent. X. The vanadium IV and vanadium V complexes of calcichrome and a spectrophotometric method based on the vanadium (IV) complex. Anal. Abstr. 21: 3355, 1971.

91. Jaraczewska, W., and M. Jakubowski. Preliminary evaluation of exposure to vanadium dust in chemical industry. Presented at XIV International Congress on Occupational Health, Madrid, 1963.

92. Jerman, L., and V. Jettmar. Polarographische Bestimmung von Vanadin in der Luft von Arbeitsräumen. Z. Ges. Hyg. 14:12–14, 1968.

93. Kang, C–C. C., E. W. Keel, and E. Solomon. Determination of traces of vanadium, iron, and nickel in petroleum oils by x-ray emission spectrography. Anal. Chem. 32:221–225, 1960.

94. Karayannis, N. M., and A. H. Corwin. Volatilization and separations of metal acetylacetonates at 115°C by hyperpressure gas chromatography. J. Chromatogr. Sci. 8:251–256, 1970.

95. Keane, J. F., and E. M. R. Fisher. Analysis of trace elements in airborne particulates by neutron activation and γ spectrometry. Atmos. Environ. 2:603–614, 1968.

96. King, R. B. Organometallic Syntheses. Vol. I. Transition-Metal Compounds. New York: Academic Press, Inc., 1965. 186 pp.

97. Kneip, T. J., M. Eisenbud, C. D. Strehlow, and P. C. Freudenthal. Airborne particulates in New York City. J. Air Pollut. Control Assoc. 20:144–149, 1970.

98. Kostromin, A. I., A. A. Akhmetov, and L. N. Orlova. Coulometric determination of manganese (II), cesium (III), and vanadium (IV). Zh. Anal. Khim. 25:195–196, 1970. (in Russian)

99. Lambert, J. P. F., R. E. Simpson, H. E. Mohr, and L. L. Hopkins, Jr. Determination of vanadium by neutron activation analysis at nanogram levels to formulate a low-vanadium diet. J. Assoc. Off. Anal. Chem. 53:1145–1150, 1970.

100. Lander, D. W., R. L. Steiner, D. H. Anderson, and R. L. Dehm. Spectrographic determination of elements in airborne dirt. Appl. Spectrosc. 25:270–275, 1971.

101. Leiritie, M., and B. Mattsson. A new approach to the standard-addition technique in atomic-absorption spectroscopy. Anal. Lett. 3:315–322, 1970.

102. Levine, E. P. Occurrence of titanium, vanadium, chromium, and sulfuric acid in the ascidian Eudistoma ritteri. Science 133:1352–1353, 1961.

103. Lewis, C. E. The biological actions of vanadium. I. Effects upon serum cholesterol levels in man. A.M.A. Arch. Ind. Health 19:419–425, 1959.

104. Lewis, C. E. The biological actions of vanadium. III. The effect of vanadium on the excretion of 5-hydroxyindolacetic acid and amino acids and the electrocardiogram of the dog. A.M.A. Arch. Ind. Health 20:455–466, 1959.

105. Lewis, C. E. The biological effects of vanadium. II. The signs and symptoms

of occupational vanadium exposure. A.M.A. Arch. Ind. Health 19:497–503, 1959.

106. Lillie, R. J. Vanadium, pp. 98–99. In Air Pollutants Affecting the Performance of Domestic Animals—A Literature Review. Agriculture Handbook No. 380. Washington, D.C.: U.S. Department of Agriculture, 1970.

107. Linstedt, K., and P. Kruger. Determination of vanadium in natural waters by neutron activation analysis. Anal. Chem. 42:113–115, 1970.

108. Livingston, H. D., and H. Smith. Estimation of vanadium in biological material by neutron activation analysis. Anal. Chem. 37:1285–1287, 1965.

109. Magyar, B. Über die Genauigkeit und Anwendbarkeit der Röntgenfluorszenz für die Bestimmung der Konzentration der Elemente Phosphor bis Uran in Lösung. Talanta 18:27–38, 1971. (summary in English)

110. Majumdar, A. K., and S. K. Bhowal. Spectrophotometric determination of vanadium with N-benzoyl-o-tolylhydroxylamine. Analyst 96:127–129, 1971.

111. Marich, K. W., P. W. Carr, W. J. Tretyl, and D. Glick. Effect of matrix material on laser-induced elemental spectral emission. Anal. Chem. 42:1775–1779, 1970.

112. Martin, G. M., E. P. Benditt, and N. Eriksen. Vanadium catalysis of the oxidation of catechol amines, dihydroxyphenylalanine and 5-hydroxyindoles. Nature 186:884–885, 1960.

113. Mascitelli–Coriandoli, E., and C. Citterio. Effects of Vanadium upon liver coenzyme A in rats. Nature 183:1527–1528, 1959.

114. Mascitelli–Coriandoli, E., and C. Citterio. Intracellular thioctic acid and coenzyme A following vanadium treatment. Nature 184:1641, 1959.

115. Mason, B. H. Principles of Geochemistry. (3rd ed.) New York: John Wiley & Sons, 1966. 329 pp.

116. McLundie, A. C., J. B. Shepherd, and D. R. A. Mobbs. Studies on the effects of various ions on enamel solubility. Arch. Oral Biol. 13:1321–1330, 1968.

117. McNay, L. M. Coal Refuse Fires, An Environmental Hazard. Washington, D.C.: Government Printing Office, 1971. 50 pp.

118. McTurk, L. C., C. H. W. Hirs, and R. E. Eckardt. Health hazards of vanadium-containing residual oil ash. Ind. Med. Surg. 25:29–36, 1956.

119. Meinke, W. W. and B. F. Scribner, Eds. First Materials Research Symposium. Trace Characterization, Chemical and Physical. Washington, D.C.: Government Printing Office, 1967. 580 pp.

120. Mitchell, R. L. Emission spectrochemical analysis. Determination of trace elements in plants and other biological materials, pp. 398–412. In J. H. Yoe and H. J. Koch, Eds. Trace Analysis. Papers Presented at a Symposium on Trace Analyses Held at the New York Academy of Medicine, New York, N.Y., November 2, 3, 4, 1955. New York: John Wiley and Sons, Inc., 1957.

121. Mitchell, W. G. Influence of pH on toxicity of vanadium in mice. Proc. Soc. Exp. Biol. Med. 84:404–405, 1953.

122. Mitchell, W. G., and E. P. Floyd. Ascorbic acid and ethylenediaminetetra-acetate (EDTA) as antidotes in experimental vanadium poisoning. Proc. Soc. Exp. Biol. Med. 85:206–208, 1954.

123. Morgan, G. B., and R. E. Homan. The Determination of Atmospheric Metals by Atomic Absorption Spectrophotometry. Oral Presentation at Pittsburgh Conference on Analytical Chemistry and Applied Spectroscopy. February 1967.

124. Morrison, G. H., Ed. Trace Analysis. Physical Methods. New York: Inter-science Publishers, Inc., 1965. 582 pp.
125. Morrow, N. L., and R. S. Brief. Elemental composition of suspended particu-late matter in metropolitan New York. Environ. Sci. Tech. 5:786–789, 1971.
126. Moshier, R. W., and R. E. Sievers. Gas Chromatography of Metal Chelates. New York: Pergamon Press, 1965. 163 pp.
127. Mountain, J. T., L. L. Delker, and H. E. Stokinger. Studies in vanadium toxi-cology: Reduction in the cystine content of rat hair. A.M.A. Arch. Ind. Hyg. Occup. Med. 8:406–411, 1953.
128. Mountain, J. T., F. R. Stockell, Jr., and H. E. Stokinger. Studies in vanadium toxicology. III. Fingernail cystine as an early indicator of metabolic changes in vanadium workers. A.M.A. Arch. Ind. Health 12:494–502, 1955.
129. Moyers, J. L., W. H. Zoller, R. A. Duce, and G. L. Hoffman. Gaseous bromine and particulate lead, vanadium and bromine in a polluted atmosphere. Environ. Sci. Technol. 6:68–71, 1972.
130. Muhler, J. C. The effect of vanadium pentoxide, fluorides, and tin compounds on the dental experience of rats. J. Dental Res. 36:787–794, 1957.
131. Myers, V. C., and H. H. Beard. Studies in the nutritional anemias of rats. II. Influence of iron plus supplements of other inorganic elements upon blood regeneration. J. Biol. Chem. 94:89–110, 1931.
132. National Academy of Sciences, National Research Council, Division of En-gineering. Trends in the Use of Vanadium. A Report of the National Materials Advisory Board. Publication NMAB-267. Springfield, Va.: Clearinghouse for Federal Scientific and Technical Information, 1970. 46 pp.
133. Natrella, M. G. Experimental Statistics, errata. National Bureau of Standards Handbook 91. Washington, D.C.: National Bureau of Standards, 1966. 3 pp.
134. Nelson, T. S., M. B. Gillis, and H. T. Peeler. Studies of the effect of vanadium on chick growth. Poult. Sci. 41:519–522, 1962.
135. Nohe, J. D., and A. J. Mitteldorf. Optimizing accuracy in emission spectro-chemical analysis. Spex Speaker 10:1–7, 1965.
136. Omang, S. H. The determination of vanadium and nickel in mineral oils by flameless graphite tube atomization. Anal. Chim. Acta 56:470–473, 1971.
137. Patrovsky, V. 2,3-Dihydroxynaphthalin als neues Reagens zur Extraktiven photometrischen Bestimmung von Eisen-, Vanadin-, Titan, und Moly-bdänspuren. Collection Czech. Commun. 35:1599–1604, 1970.
138. Patty, F. A. The mode of entry and action of toxic materials, pp. 143–172. In F. A. Patty, Ed. Industrial Hygiene and Toxicology. Vol. I. New York: Interscience Publishers, Inc., 1958.
139. Pazynich, V. M. Maximum permissible concentration of vanadium pentoxide in the atmosphere. Hyg. Sanit. 31:6–12, 1966.
140. Perry, H. M., Jr., P. L. Schwartz, and B. M. Sahagian. Effect of transition metals and of metal binding antihypertensive agents on tryptamine oxidase and dopa decarboxylase. Proc. Soc. Exp. Biol. Med. 130:273–277, 1969.
141. Perry, H. M., Jr., S. Teitlebaum, and P. L. Schwartz. Effect of antihyper-tensive agents on amino acid decarboxylation and amine oxidation. Fed. Proc. 14:113–114, 1955.
142. Peterson, P. J. Unusual accumulations of elements by plants and animals. Sci. Prog. 59:505–526, 1971.
143. Pillay, K. K. S., and C. C. Thomas, Jr. Determination of the trace element

levels in atmospheric pollutants by neutron activation analysis. J. Radioanal. Chem. 7:107–118, 1971.

144. Prince, A. L. Trace element delivering capacity of 10 New Jersey soil types as measured by spectrographic analyses of soils and mature corn leaves. Soil Sci. 84:413–418, 1957.

145. Radford, H. D., and R. C. Rigg. New way to desulfurize resids. Hydrocarbon Process. 49:187–191, 1970.

146. Rahn, K. A. Sources of Trace Elements in Aerosols: An Approach to Clean Air. U.S. Atomic Energy Commission Contract No. COO-1705-9. Springfield, Va.: U.S. Department of Commerce, National Technical Information Service, 1971. 309 pp.

147. Rahn, K. A., R. Dams, J. A. Robbins, and J. W. Winchester. Diurnal variations of aerosol trace element concentrations as determined by nondestructive neutron activation analysis. Atmos. Environ. 5:413–422, 1971.

148. Ralson, H. R., and E. S. Sato. Sodium removal as an aid to neutron activation analysis. Anal. Chem. 43:129–131, 1971.

149. Rhoads, H. U. Analysis of Atmospheric Dust by Electron Optics. Public Health Service Grant 2R01 AP 00372 active for the period of February 1, 1966 to January 31, 1970. Terminal Report. 27 pp. (unpublished)

150. Rigdon, L. P., and J. E. Harrar. Determination of vanadium by controlled-potential coulometry. Anal. Chem. 41:1673–1675, 1969.

151. Ritchie, P. D., Ed. Vinyl and Allied Polymers, p. 73. London: Iliffe Books, Ltd., 1968.

152. Robertson, D. E. Role of contamination in trace element analysis of sea water. Anal. Chem. 40:1067–1072, 1968.

153. Romoser, G. L., W. A. Dudley, L. J. Machlin, and L. Loveless. Toxicity of vanadium and chromium for the growing chick. Poult. Sci. 40:1171–1173, 1961.

154. Roshchin, I. V. Toxicology of vanadium compounds used in modern industry. Gig. Sanit. 32:26–32, 1967.

155. Roshchin, I. V., A. V. Il'nitskaia, L. A. Lutsenko, and L. V. Zhidkova. Effect on organism of vanadium trioxide. Fed. Proc. 24:611–613, July–Aug. 1965.

156. Rostoker, W. The Metallurgy of Vanadium. New York: John Wiley and Sons, 1958. 185 pp.

157. Schlain, D., C. B. Kenahan, and W. L. Acherman. Corrosion behavior of high-purity vanadium. Corrosion 16:70t–72t, 1960.

158. Schroeder, H. A. Air Quality Monographs. Monograph 70-13. Vanadium. Washington, D.C.: American Petroleum Institute, 1970. 32 pp.

159. Schroeder, H. A. Mechanisms of Hypertension, with a Consideration of Atherosclerosis. American Lecture Series No. 305; American Lectures in Metabolism. Springfield, Ill.: Charles C Thomas, 1957. 362 pp.

160. Schroeder, H. A. Possible relationships between trace metals and chronic disease, pp. 59–67. In M. J. Severn and L. A. Johnson, Eds. Metal Binding in Medicine; Proceedings of a Symposium Held in Philadelphia, May 6–8, 1959. Philadelphia: Lippincott & Co., 1960.

161. Schroeder, H. A., J. J. Balassa, and I. H. Tipton. Abnormal trace metals in man—vanadium. J. Chron. Dis. 16:1047–1071, 1963.

162. Schumann–Vogt, B. Vanadinbedingte Gesundheitsschäden in der Industrie. Zentralbl. Arbeitsmed. 19:33–39, 1969.

163. Schwarz, K., and D. B. Milne. Growth effects of vanadium in the rat. Science 174:426–428, 1971.

164. Second Annual Report of the Council on Environmental Quality. Washington, D.C.: Government Printing Office, 1971. 360 pp.

165. Seidell, A. Solubilities of Inorganic and Metal Organic Compounds. A compilation of Quantitative Solubility Data from the Periodical Literature, p. 1569. 3rd ed. Vol. I. New York: D. Van Nostrand Co., 1940.

166. Shatkay, A. A critical analysis of the method of successive dilutions in photometry. Anal. Chim. Acta 52:547–550, 1970.

167. Shatkay, A. Photometric determination of substances in presence of strongly interfering unknown media. Anal. Chem. 40:2097–2106, 1968.

168. Sierra, F., C. Sanchez–Pedrino, T. Perez–Ruiz, and C. Martinez–Lozano. Amperometric determination of vanadates. An. Quim. 66:479–486, 1970.

169. Singh, D., and S. Sharma. Amperometric permanganometric estimations at low concentrations in stirred solutions. Indian J. Chem. 8(2):192, 1970.

170. Siroki, M., and C. Djordjevic. Spectrophotometric determination of vanadium with 4-(2-pyridylazo)resorcinaol by extracting of tetraphenyl-phosphonium and arsonium salts. Anal. Chim. Acta 57:301–310, 1971.

171. Sjöberg, S. G. Vanadium bronchitis after cleaning of oil burning steam boilers. Nord. Hyg. Tidskr. 35(3–4):45–57, 1954. (in Swedish; summary in English)

172. Sjöberg, S–G. Vanadium dust, chronic bronchitis and possible risk of emphysema: A follow-up investigation of workers at a vanadium factory. Acta Med. Scand. 154:381–386, 1956.

173. Sjöberg, S. G. Vanadium pentoxide dust. A clinical and experimental investigation on its effect after inhalation. Acta Med. Scand. 138 (Suppl. 238), 1950. 188 pp.

174. Sjöberg, S. G., and K–G. Rigner. Skin, eye, and respiratory tract symptoms associated with cleaning of oil-fired boilers; an investigation with special reference to vanadium in blood and urine. Nord. Hyg. Tidskr. 37(9–10):217–228, 1956. (in Swedish; summary in English)

175. Snyder, F., and W. E. Cornatzer. Vanadium inhibition of phospholipid synthesis and sulphydryl activity in rat liver. Nature 182:462, 1958.

176. Somerville, J., and B. Davies. Effect of vanadium on serum cholesterol. Amer. Heart J. 54:54–56, 1962.

177. Söremark, R. Vanadium in some biological specimens. J. Nutr. 92:183–190, 1967.

178. Söremark., and S. Üllberg. Distribution and kinetics of $^{48}V_2O_5$ in mice, pp. 103–114. In Use of Radioisotopes in Animal Biology and the Medical Sciences. Proceedings of a Conference Held in Mexico City, 21 Nov.–1 Dec., 1961. Vol. 2. New York: Academic Press, 1962.

179. Söremark, R., S. Ullberg, and L–E. Appelgren. Autoradiographic localization of V-48-labelled vanadium pentoxide in developing teeth and bones of rats. Acta Odont. Scand. 20:225–232, 1962.

180. Steel, J., and J. T. Sanderson. Toxic constituents of welding fumes. Ann. Occup. Hyg. 9:103–111, 1966.

181. Stocks, P. On the relations between atmospheric pollution in urban and rural localities and mortality from cancer, bronchitis, and pneumonia, with particular to 3,4-benzopyrene, beryllium, molybdenum, vanadium and arsenic. Brit. J. Cancer 14:397–418, 1960.

182. Stokinger, H. E. Organic, beryllium and vanadium dusts. A review. A.M.A. Arch. Ind. Health 12:675–677, 1955.

183. Stokinger, H. E. Vanadium, V, pp. 1171–1181. In F. A. Patty, Ed. Industrial Hygiene and Toxicology. Vol. II. Toxicology. (2nd rev. ed.) New York: Interscience Publishers, 1963.

184. Strain, W. H. Effects of some minor elements on animals and people, p. 22. Presented to Symposium, "Geochemical Evolution—The First Five Billion Years," American Association for the Advancement of Science, Denver, Colorado, December 29, 1961.

185. Strain, W. H., A. M. Dutton, A. D'Angelo, N. C. Plumstead, and G. H. Ramsey. Progress Report on Acceleration of Burn and Wound Healing with Trace Elements (Period: January 1, 1955 to December 31, 1956). Rochester, N.Y.: University of Rochester, 1957. 27 pp.

186. Strasia, C. A. Vanadium: Essentiality and Toxicity in the Laboratory Rat. Ph.D. Thesis. Purdue University, 1971. Dissertation Abstr. 32:646-B, 1971.

187. Sykes, C., and H. T. Shirley. Scaling of heat-resisting steels. Influence of combustible sulfur and oil–fuel ash constituents, pp. 153–169. In High-Temperature Steels and Alloys for Gas Turbines. Iron and Steel Institute (London), Special Report No. 43, July 1952.

188. Symanski, [J.]. Gewerbliche Vanadinschädigungen, ihre Enstehung und Symptomatologie. Arch. Gewerbepath. Gewerbehyg. 9:295–313, 1939.

189. Symposium on Standard Methods for the Analysis of Air Pollutants. Sponsored by the Intersociety Committee on Methods of Air Sampling and Analysis. M. Katz, Presiding. Oral Presentation at Pittsburgh Conference on Analytical Chemistry and Applied Spectroscopy. March 1972.

190. Talvitie, N. A. Colorimetric determination of vanadium with 8-quinolinol application to biological materials. Anal. Chem. 25:604–607, 1953.

191. Tank, G., and C. A. Storvick. Effect of naturally occurring selenium and vanadium on dental caries. J. Dental Res. 39:473–488, 1960.

192. Taylor, M. L. Gas liquid chromatography of trace elements, pp. 363–389. In W. Mertz and W. E. Cornatzer, Eds. Newer Trace Elements in Nutrition. New York: Marcel Dekker, 1971.

193. Tebrock, H. E., and W. Machle. Exposure to europium-activated yttrium orthovanadate: A cathodoluminescent phosphor. J. Occup. Med. 10:692–696, 1968.

194. Thompson, R. J., G. B. Morgan, and L. J. Purdue. Analysis of selected elements in atmospheric particulate matter by atomic absorption. Atomic Absorption Newslett. 9:53–57, 1970.

195. Tipton, I. H., and M. J. Cook. Trace elements in human tissue. II. Adult subjects from the United States. Health Phys. 9:103–107, 1963.

196. Tipton, I. H., M. J. Cook, R. L. Steiner, C. A. Boyle, H. M. Perry, Jr., and H. A. Schroeder. Trace elements in human tissue. I. Methods. Health Phys. 9:89–91, 1963.

197. Tipton, I. H., H. A. Schroeder, H. M. Perry, Jr., and M. J. Cook. Trace elements in human tissue. III. Subjects from Africa, the Near and Far East and Europe. Health Phys. 11:403–451, 1965.

198. Tipton, I. H., and J. J. Shafer. Statistical analysis of lung trace element levels. Arch. Environ. Health 8:58–67, 1964.

199. Underwood, E. J. Trace Elements in Human and Animal Nutrition. (3rd ed.) New York: Academic Press Inc., 1971. 543 pp.

200. U.S. Department of Health, Education, and Welfare, Public Health Service, Consumer Protection and Environmental Health Service. Air Quality Data from the National Air Surveillance Networks and Contributing State and Local Networks. 1966 Edition. NAPCA Publication APTD 68-9. Durham, N.C.: National Air Pollution Control Administration, 1968. 157 pp.

201. Valberg, L. S., and J. M. Holt. Detection of vanadium in normal human erythrocytes. Life Sci. 3:1263–1265, 1964.

202. Vinogradov, A. P. The Geochemistry of Rate and Dispersed Elements in Soil. (2nd ed.) New York: Consultants Bureau, Inc., 1959. 209 pp.

203. Wahler, W. Mechanical and Chemical Dressing of Minerals and Rocks for Geochemical Trace Analysis. NASA Technical Translation F-10,718. Washington, D.C.: National Aeronautics and Space Administration, 1967. 17 pp.

204. Watanabe, H., H. Murayama, and S. Yamaoka. Some clinical findings on vanadium workers. Jap. J. Ind. Health 8(7):23–27, 1966.

205. Weast, R. C., Ed. CRC Handbook of Chemistry and Physics. (51st ed.) Cleveland, Ohio: Chemical Rubber Co., 1970. 2400 pp.

206. Welz, B., and E. Wiedeking. Bestimmung von Spurenelementen in Serum und Urin mit flammenloser Atomisierung. Fresinius' Z. Anal. Chem. 252: 111–117, 1970.

207. Williams, N. Vanadium poisoning from cleaning oil-fired boilers. Brit. J. Ind. Med. 9:50–55, 1952.

208. Winter, J. A. Interlaboratory Quality Control in the Federal Water Quality Administration. Oral Presentation, American Chemical Society, Chicago, Ill., Sept. 1970.

209. Wyers, H. Some toxic effects of vanadium pentoxide. Brit. J. Ind. Med. 3:177–182, 1946.

210. Yoe, J. H., and H. J. Koch, Eds. Trace Analysis. Papers Presented at a Symposium on Trace Analyses Held at the New York Academy of Medicine, New York, N.Y., November 2,3,4, 1955. New York: John Wiley and Sons, Inc., 1957. 672 pp.

211. Yurachek, J. P., G. G. Clemena, and W. W. Harrison. Analysis of human hair by spark source mass spectrometry. Anal. Chem. 41:1666–1668, 1969.

212. Zadumina, E. A., and A. I. Cherkesov. Photometric determination of vanadium with picraminazo-N. Izv. Vyssh. Ucheb. Zaved. Khim. Khim. Tekhnol. 12:1483–1486, 1969. (in Russian)

213. Zenz, C., J. P. Bartlett, and W. H. Thiede. Acute vanadium pentoxide intoxication. Arch. Environ. Health 5:542–546, 1962.

214. Zenz, C., and B. A. Berg. Human responses to controlled vanadium pentoxide exposure. Arch. Environ. Health 14:709–712, 1967.

215. Zief, M. A. Chemical purebreds. Ind. Res. 13:36–39, 1971.

216. Zoller, W. H., and G. E. Gordon. Instrumental neutron activation analysis of atmospheric pollutants utilizing Ge(Li) x-ray detectors. Anal. Chem. 42:257–265, 1970.

217. Zoller, W. H., G. E. Gordon, E. S. Gladney, and A. G. Jones. The sources and distribution of vanadium in the atmosphere. Amer. Chem. Soc., Div. Water, Air Waste Chem., Gen. Pap. 11:159–165, 1971.

Index

111

ISBN 0-30